CONFIDENCE INTERVALS ON VARIANCE COMPONENTS

STATISTICS: Textbooks and Monographs

A Series Edited by

D. B. Owen, Coordinating Editor
Department of Statistics
Southern Methodist University
Dallas, Texas

R. G. Cornell, Associate Editor
for Biostatistics
University of Michigan

W. J. Kennedy, Associate Editor
for Statistical Computing
Iowa State University

A. M. Kshirsagar, Associate Editor
for Multivariate Analysis and
Experimental Design
University of Michigan

E. G. Schilling, Associate Editor
for Statistical Quality Control
Rochester Institute of Technology

Additional Volumes in Preparation

CONFIDENCE INTERVALS ON VARIANCE COMPONENTS

Arizona State University
Tempe, Arizona

FRANKLIN A. GRAYBILL

Colorado State University
Fort Collins, Colorado

CRC Press
Taylor & Francis Group
Boca Raton London New York

CRC Press is an imprint of the
Taylor & Francis Group, an **informa** business

CRC Press
Taylor & Francis Group
6000 Broken Sound Parkway NW, Suite 300
Boca Raton, FL 33487-2742

First issued in paperback 2019

© 1992 by Taylor & Francis Group, LLC
CRC Press is an imprint of Taylor & Francis Group, an Informa business

No claim to original U.S. Government works

ISBN-13: 978-0-8247-8644-1 (hbk)
ISBN-13: 978-0-367-40282-2 (pbk)

Visit the Taylor & Francis Web site at
http://www.taylorandfrancis.com

and the CRC Press Web site at
http://www.crcpress.com

Library of Congress Cataloging-in-Publication Data

Burdick, Richard K.
    Confidence intervals on variance components / Richard K. Burdick,
Franklin A. Graybill.
        p.   cm. -- (Statistics, textbooks and monographs ; 127)
    Includes bibliographical references and index.
    ISBN 0-8247-8644-0 (alk. paper)
    1. Confidence intervals.   2. Analysis of variance.
3. Experimental design.   I. Graybill, Franklin A.   II. Title.
III. Series: Statistics, textbooks and monographs ; v. 127.
QA276.74.B87   1992
519.5'38 -- dc20   92-3322
                                                                        CIP

Preface

Many applications in a variety of fields require the measurement of variance. To effectively understand these measurements, decision makers require both point and interval estimates. The purpose of this book is to present methods for constructing confidence interval estimates on measures of variation. The book is intended for practitioners who should be using these methods in their fields of application. Researchers who are primarily interested in the theoretical development of these methods will also find sufficient references to help them in their work. Because variance measures are of interest in many fields, articles concerning confidence intervals on variances are spread throughout the literature. In addition, much of this research has only recently been completed, and most linear model and experimental design textbooks are lacking in its coverage. By summarizing this research, it is our hope this book will serve the needs of both practicing and theoretical statisticians.

This book contains methods for constructing confidence intervals on individual variance components, linear combinations of variance components, and ratios of variance components for a variety of experimental designs. These designs include both crossed and nested factors, with both balanced and unbalanced data sets. The only prerequisite is a course in elementary statistical analysis and a basic understanding of statistical inference. All of the formulas presented in the book are easily computed by hand and simple computer codes can be written for multiple applications. Numerical examples are used to demonstrate their application. Theoretical development is minimal although appropriate references are cited for readers who desire a more theoretical understanding of the material.

Chapter 1 provides some history of variance component estimation and defines more clearly the focus of the book. Chapter 2 presents basic results and terminology that are used in the book. General methods for constructing confidence intervals on functions of variance components are presented in Chapter 3. These results are used to construct confidence intervals for all the designs that are considered in later chapters. The designs considered in the text include the one-fold nested random model (Chapter 4), the two-fold and $(Q\text{-}1)$-fold nested random models (Chapter 5), crossed random models (Chapter 6), and mixed models (Chapter 7).

We hope this book will make more analysts aware of the methods that can be used for constructing confidence intervals on variance components. We have attempted to explain these methods in a manner that encourages investigators with only a modest statistical background to use them. Because of the wide interest in variance measures, we believe these simple methods can successfully aid decision makers in many fields. Finally, we hope that others will contribute to this body of research by developing new and better methodology.

Richard K. Burdick
Franklin A. Graybill

Contents

1
Introduction

1.1 BACKGROUND

In this section we define three general types of models that are discussed in this book. In the simplest case consider a sample from a normal population with mean μ and variance σ^2 where one is interested in inferences on μ and σ^2. To generalize, consider sampling from k populations with means and variances μ_i and σ_i^2, respectively. If the σ_i^2 are equal and the interest is to make statistical inferences about the μ_i, the model is referred to as a fixed effects model (also called a design model or Eisenhart's Model I). If the μ_i are equal and the interest is in making inferences about the σ_i^2, the model is referred to as a random effects model (also called a components-of-variance model, or Eisenhart's Model II). If the interest is in making inferences about the μ_i and/or the σ_i^2 when they are different for each population, the model is referred to as a mixed model (or Eisenhart's Model III). We shall discuss each model briefly.

1.1.1 Fixed Effects Model

The simplest fixed effects model is

$$Y_{ij} = \mu + \alpha_i + E_{ij} \qquad i = 1, \ldots, I; \quad j = 1, \ldots, J \ (1.1.1)$$

There are I populations and a random sample of size J is selected from each where Y_{ij} is the observed jth sample value from the ith population. The quantities μ and α_i are unobservable fixed constants called parameters. The mean of the ith population is $\mu + \alpha_i$ and the variance of the ith population is σ_E^2. Thus, the E_{ij} have means zero and variances σ_E^2. The objective in this model is to make inferences about certain relationships among the α_i. A general fixed effects model is

$$Y_{ij\ldots p} = \mu + \alpha_i + \beta_j + \gamma_{ij} + \cdots + E_{ij\ldots p} \qquad (1.1.2)$$

where $Y_{ij\ldots p}$ is an observed value, the quantities μ, α_i, β_j, and γ_{ij} are unobservable constant parameters and $E_{ij\ldots p}$ is an unobservable random variable with mean zero and variance σ_E^2. In fixed models of this type, the objective is generally to make statistical inferences about the unknown parameters μ, α_i, β_j, and γ_{ij}. The following example illustrates a fixed effects model.

Example 1.1.1 An investigator wants to measure the effect on the yield of a certain variety of corn of two different chemical compounds and four different methods of applying these compounds. Suppose α_i is the effect of the ith chemical ($i = 1,2$) and β_j is the effect of the jth method of applying the compounds ($j = 1,2,3,4$). If the investigator assumes that the effects are additive, the following model is assumed:

$$\begin{aligned}
Y_{11} &= \mu + \alpha_1 + \beta_1 + E_{11} \\
Y_{12} &= \mu + \alpha_1 + \beta_2 + E_{12} \\
Y_{13} &= \mu + \alpha_1 + \beta_3 + E_{13} \\
Y_{14} &= \mu + \alpha_1 + \beta_4 + E_{14} \\
Y_{21} &= \mu + \alpha_2 + \beta_1 + E_{21} \\
Y_{22} &= \mu + \alpha_2 + \beta_2 + E_{22} \\
Y_{23} &= \mu + \alpha_2 + \beta_3 + E_{23} \\
Y_{24} &= \mu + \alpha_2 + \beta_4 + E_{24}
\end{aligned}$$

This model is written in a more compact form as

$$Y_{ij} = \mu + \alpha_i + \beta_j + E_{ij} \qquad i = 1,2; \quad j = 1,2,3,4 \qquad (1.1.3)$$

where Y_{ij} is the observable yield of corn on a plot of ground that received the ith chemical applied by the jth method. The investigator assumes the observable random variable Y_{ij} is equal to a constant μ, the average yield when no chemical is applied and hence no method of application is used, plus α_i, the effect due to the ith chemical, plus β_j, the effect due to the jth method of application, plus a random error E_{ij} due to all the uncontrolled factors, such as differences of fertility among the plots. The experimenter may desire to test or estimate estimable functions of the parameters μ, α_i, and β_j.

1.1.2 Components-of-Variance Model

A general components-of-variance model is

$$Y_{ij\ldots p} = \mu + A_i + B_j + \cdots + E_{ij\ldots p} \qquad (1.1.4)$$

where $A_i, B_j, \ldots$, and $E_{ij\ldots p}$ are random variables with means of zero and variances $\sigma_A^2, \sigma_B^2, \ldots, \sigma_E^2$, respectively. The objective in model (1.1.4) is to make statistical inferences about certain functions of the variances. We illustrate with an example.

Example 1.1.2 In examining the nitrogen content of the foliage in a large orchard, leaves from trees are collected, and the nitrogen content of the leaves is measured. There are two major sources of variation: the variation of the leaves on a tree, and the variation among trees in the orchard. The objective is to measure these two variances. However, it is impossible to measure the "true" nitrogen content of the foliage of a tree without stripping all the leaves from the tree. That is, each tree in the orchard has a "true" nitrogen content of its leaves although this true value can be obtained only by measuring the nitrogen content of every leaf on the tree. The distribution of "true" nitrogen contents of all trees in the orchard could be found by determining the "true" nitrogen content of each tree in the orchard by measuring each leaf on each tree. One problem is to determine σ_T^2, the variance of the "true" nitrogen contents of the trees. The difficulty is that we cannot observe the "true" nitrogen content of any tree. Thus, we are in the position of wanting to determine (estimate) the variance of a random variable T and not being able to observe any values of the random variable. To accomplish our objective we will build a theoretical model, select some trees at random from the orchard, select some leaves at random from each of these trees, measure the nitrogen content of the selected leaves, and use these observed values with the theoretical model to estimate σ_T^2. We consider two basic populations. One population is the set of numbers denoted by T, that represent the "true" nitrogen content of trees: we let μ and σ_T^2 be the mean and variance of T. Another population (actually a family of populations) is obtained by considering the leaves on each tree as a population. For a given tree the nitrogen content of the leaves is a population with mean equal to t_o, say, and this value is the "true" nitrogen content of that tree. We assume that the variance of the nitrogen content of the leaves on the given tree is σ_L^2 and is the same for all trees. The sampling scheme we use is as follows:

1. A tree is selected at random from the orchard, and we assume the "true" unobservable nitrogen content is T_1^*.
2. J leaves are selected at random from this tree, and we measure the nitrogen content of each leaf and denote the J observable values by $Y_{11}, Y_{12}, \ldots, Y_{1J}$.
3. We write the model for the first tree selected as

$$Y_{1j} = \mu + (T_1^* - \mu) + (Y_{1j} - T_1^*) \qquad j = 1, 2, \ldots, J$$

where μ is the mean of the "true" nitrogen content of all leaves on all trees in the orchard.

4. In all, we draw I trees at random and assume that the "true" nitrogen contents are respectively $T_1^*, T_2^*, \ldots, T_I^*$. We draw J leaves at random from each tree and denote the observable nitrogen content of the jth leaf from the ith tree by Y_{ij}. We write the model as

$$Y_{ij} = \mu + (T_i^* - \mu) + (Y_{ij} - T_i^*)$$

or as

$$Y_{ij} = \mu + T_i + L_{ij} \qquad i = 1, 2, \ldots, I; \quad j = 1, 2, \ldots, J$$

where $T_i = T_i^* - \mu$ and $L_{ij} = Y_{ij} - T_i^*$. The term T_i is the effect of the ith tree, where $\text{Exp}[T_i] = 0$, $\text{Var}[T_i] = \sigma_T^2$, and Exp and Var denote the expected value and variance operators, respectively. The term L_{ij} is the effect of the jth leaf from the ith tree, and hence $\text{Exp}[L_{ij}] = 0$, and $\text{Var}[L_{ij}] = \sigma_L^2$ for each i and j. We also assume that $T_{i'}$ and L_{ij} are uncorrelated for all i, i', and j. With the properties mentioned, it follows that

$$\text{Var}[Y_{ij}] = \text{Var}[T_i] + \text{Var}[L_{ij}]$$

or

$$\sigma_Y^2 = \sigma_T^2 + \sigma_L^2$$

where σ_T^2 and σ_L^2 are "components" of the variance of the observable Y_{ij}. We may also want to assume distributional properties (such as normality) for the T_i and L_{ij}.

Now we abstract the important concepts of Example 1.1.2 and define the variance components model. Let Y_{ij} be an observable random variable with a structure

$$Y_{ij} = \mu + A_i + E_{ij} \qquad i = 1, ..., I; \quad j = 1, ..., J \ (1.1.5)$$

where μ is an unknown parameter, and A_i and E_{ij} are uncorrelated and unobservable random variables with zero means and variances σ_A^2 and σ_E^2, respectively. These specifications define a variance component (or components-of-variance) model. Sometimes there will be additional distributional assumptions for the random variables A_i and E_{ij}.

1.1.3 Mixed Model

In a mixed model some of the quantities in the model (1.1.2) are random variables and some are fixed parameters. The objective in these models is to make statistical inferences about the fixed parameters and the variances of the random variables. Most of this book is concerned with the components-of-variance model, although Chapter 7 discusses some mixed models.

1.1.4 Some History of Components-of-Variance Models

Of the three models discussed in this chapter, the fixed effects model is by far the most developed. It is interesting that R. A. Fisher (1924) developed a procedure for analyzing the fixed model and called the procedure "the analysis of variance." The procedure included optimum methods for point estimation, confidence intervals, and hypothesis tests. In a classical paper S. Kolodziejczyk (1935) gave a general theory of the linear model that included the fixed effects model. During the 1930s, 1940s, and 1950s a tremendous amount of work extended and generalized this model. The components-of-variance model received much less attention during this period in part because it was not so amenable to optimum procedures and exact confidence intervals and tests. One of the early papers that attempted approximate procedures was by Smith (1936). Two of the important early papers were by Yates and Zacopanay (1935) and Daniels (1939). By 1951 enough work on components-of-variance models had been done for Crump (1951) to write a paper entitled, "The present status of variance components." Most of the papers on variance components during the 1930s and 1940s concerned point estimation. During the late 1940s and throughout the 1950s much work was done on procedures for obtaining approximate confidence intervals on linear combinations of variances. This work included contributions by Satterthwaite (1941, 1946), Green (1954), Welch (1956), Huitson (1955), Bulmer (1957), and Moriguti (1955). Most of the work on variance component models during this period was based on the analysis of variance as if the model was a fixed effects model. Graybill and Hultquist (1961) gave conditions when this procedure led to optimum point estimation of linear combinations of variance components. During the 1960s and 1970s point estimation of variance components received a great deal of attention. Searle (1971) provides a summary of these and other activities. Papers concerning confidence intervals began appearing in the late 1970s. Burdick and Graybill (1988) reviewed this research and reported the present status of confidence intervals for functions of variance components. Most of the results for confidence intervals on functions of variance components have considered balanced components-of-variance models. More recently results for unbalanced models have been published. For readers who want to study all aspects of components-of-variance models, fairly complete bibliographies are provided by Sahai (1979), Khuri and Sahai (1985), and Sahai, Khuri, and Kapadia (1985). Searle (1988) reviews some of the history and results on unbalanced and mixed models.

1.2 SCOPE OF BOOK

1.2.1 Univariate Models

We consider only models with a single response variable. An overview of some of the work that has been done with multivariate response variables is provided by Anderson (1985). We address problems associated with both random and mixed models. We consider both balanced and unbalanced models.

1.2.2 Interval Estimation

The major focus of this book is on interval estimation. More specifically, we present results only on confidence intervals. Bross (1950), Venables and James (1978), and Wild (1981) have studied intervals on variance components based on fiducial distributions. Box and Tiao (1973) and Broemeling (1985) address issues concerning variance components from a Bayesian perspective. Applications of tolerance intervals for random one-fold nested designs are presented by Mee and Owen (1983), Mee (1984), Limam and Thomas (1988), and Wang (1988a).

1.2.3 Normal Theory

The methods presented in this book are based on standard normal theory assumptions. In particular, all random errors are jointly independent normal random variables with means of zero and constant variances. Some references are given in the text to papers that consider models under less restrictive assumptions.

There is a large amount of literature that considers interval estimation of variance components under non-normality. A bibliography of some of this work is provided by Singhal, Tiwari, and Sahai (1988). Additional papers of interest include Spjøtvoll (1967), Arvesen (1969), Arvesen and Schmitz (1970), Arvesen and Layard (1975), Prasad and Rao (1988), and Aastveit (1990). In situations where non-normality is

suspected, the investigator should determine whether variance measures are informative before applying any of the robust procedures
proposed in these papers. An alternative to using a robust procedure is
to transform the data to make it more normal and then apply the
methods discussed in this book. Transformations of this type are discussed in Solomon (1985).

Although based on normal theory, the intervals presented in this
book are not the large-sample normal theory intervals recommended in
papers concerning point estimation of variance components. The performance of these large-sample intervals are sometimes uncertain with
small samples. In contrast, the methods presented in this book work
well for all sample sizes. The intervals are generally approximate, but
in most cases maintain the stated confidence level.

1.2.4 Computationally Convenient Procedures

Another feature of the intervals presented in this book is that all required computations can be easily performed with a hand-held calcu-
lator. An obvious advantage of this feature is that all of the formulas
can be programmed for a computer using simple code. It is our belief
that this feature will encourage use of the methods by all investigators
regardless of their level of statistical and computational sophistication.

2
General Concepts

2.1 INTRODUCTION

This chapter presents an overview of concepts and results that are used throughout the book. Section 2.2 contains an introduction to random effects models and describes functions of variance components to be considered. The next two sections present some basic notions concerning confidence intervals. Section 2.3 provides a brief discussion of statistical inference with special emphasis on the advantages of confidence intervals and Section 2.4 discusses simultaneous confidence intervals. Section 2.5 describes the notation used in the book and illustrates the use of the tables contained in Appendix A. These tables are used in Sections 2.6 and 2.7 to provide confidence intervals on variances for one and two normal populations.

2.2 VARIANCE COMPONENT MODELS

The focus of this book concerns variance component models. Variance components are associated with random effects and appear in both random and mixed models. Random effects contain factor levels that have been randomly selected from a population of factor levels. The focus of any inference is on this population and not on the sampled factor levels. Variance components represent variances of populations of factor levels and are of interest in any investigation that contains random effects. Two examples are presented to illustrate the nature of problems addressed with variance component models.

Example 2.2.1   In examining the nitrogen content of the foliage in a large orchard, there are two major sources of variation. These sources are the variance of nitrogen content for the leaves on a given tree (denoted σ_E^2) and the variance among the nitrogen contents of the trees in the orchard (denoted by σ_A^2). An investigator desires to measure these two variances as well as other functions of σ_A^2 and σ_E^2. However, in order to measure the nitrogen content of a leaf, it must be stripped from a tree and destroyed. Since one would not want to strip all leaves from every tree in the orchard, the investigator selects a random sample of trees from the orchard and a random sample of leaves from each sampled tree. The focus of the investigation is not on the sampled trees, but rather on the population of trees in the orchard. The observed values of nitrogen content for the sampled leaves are used with a theoretical variance component model to estimate σ_A^2 and σ_E^2. This model is

$$Y_{ij} = \mu + A_i + E_{ij} \qquad i = 1, \ldots, I; \quad j = 1, \ldots, J \qquad (2.2.1)$$

where Y_{ij} is the observed nitrogen content for the jth leaf from the ith tree, μ is a constant that represents the average nitrogen content for all leaves in the orchard, A_i is a random variable with mean zero and variance σ_A^2, and E_{ij} is a random variable with mean zero and variance σ_E^2. The parameters σ_A^2 and σ_E^2 are the variance components. By observing Y_{ij} for the selected leaves on selected trees, one can estimate μ, σ_A^2, and σ_E^2. In addition, confidence intervals and tests of hypotheses concerning these parameters can be obtained. Although the theoretical model in (2.2.1) assumes samples are taken from infinite populations, the model works well in the more realistic situation where samples are taken from large finite populations.

The starting point for any analysis concerning a variance component model is the *analysis of variance* (ANOVA) table. The ANOVA table for model (2.2.1) is displayed in Table 2.2.1.

Table 2.2.1 Analysis of Variance for Model (2.2.1)

Source of variation (SV)	Degrees of freedom (DF)	Sum of squares (SS)	Mean square (MS)	Expected mean square (EMS)
Factor A (Among Trees)	$n_1 =$ $I - 1$	$SS1 =$ $\Sigma_i \Sigma_j (\bar{Y}_{i*} - \bar{Y}_{**})^2$	$S_1^2 = SS1/n_1$	$\theta_1 = \sigma_E^2 + J\sigma_A^2$
Error (Among Leaves in Trees)	$n_2 =$ $I(J - 1)$	$SS2 =$ $\Sigma_i \Sigma_j (Y_{ij} - \bar{Y}_{i*})^2$	$S_2^2 = SS2/n_2$	$\theta_2 = \sigma_E^2$
Total	$IJ - 1$	$SST =$ $\Sigma_i \Sigma_j (Y_{ij} - \bar{Y}_{**})^2$		

The notation used in this table is described more fully in Section 2.5. Formulas for making inferences on σ_A^2 and σ_E^2 are derived using S_1^2 and S_2^2. In addition to confidence intervals on the components σ_A^2 and σ_E^2, one might want confidence intervals on the total variance $\text{Var}(Y_{ij}) = \sigma_A^2 + \sigma_E^2 = \gamma$, and on the ratios σ_E^2/σ_A^2, σ_E^2/γ, and σ_A^2/γ.

The model in (2.2.1) is called the (balanced) one-fold nested random model. It is discussed in detail in Chapter 4. The next example illustrates the two-factor crossed random model.

Example 2.2.2 A quality control manager of a company that makes window screens wants to study the sources of variability in the length of the screens that are produced. Operators use special machines to cut the frames that form the sides of the screens and the manager wants to study the contribution to the variability of the final product that is due to operators, machines, and the operator-machine interaction. She randomly selects three operators from all operators in the company who use the machines and has each operator make two screens on each of four randomly selected machines. An observation Y_{ijk} is the length of the frame of the kth screen made by the ith operator using the jth machine. The theoretical model used to represent Y_{ijk} is

$$Y_{ijk} = \mu + A_i + B_j + (AB)_{ij} + E_{ijk}$$
$$i = 1, ..., I; \quad j = 1, ..., J; \quad k = 1, ..., K; \quad (2.2.2)$$

Table 2.2.2 Analysis of Variance for Model (2.2.2)

SV	DF	SS	MS	EMS
Factor A (Operators)	$n_1 = I - 1$	$SS1 = \sum_i \sum_j \sum_k (\bar{Y}_{i**} - \bar{Y}_{***})^2$	S_1^2	$\theta_1 = \sigma_E^2 + K\sigma_{AB}^2 + KJ\sigma_A^2$
Factor B (Machine)	$n_2 = J - 1$	$SS2 = \sum_i \sum_j \sum_k (\bar{Y}_{*j*} - \bar{Y}_{***})^2$	S_2^2	$\theta_2 = \sigma_E^2 + K\sigma_{AB}^2 + KI\sigma_B^2$
Interaction	$n_3 = (I - 1)(J - 1)$	$SS3 = \sum_i \sum_j \sum_k (\bar{Y}_{ij*} - \bar{Y}_{i**} - \bar{Y}_{*j*} + \bar{Y}_{***})^2$	S_3^2	$\theta_3 = \sigma_E^2 + K\sigma_{AB}^2$
Error	$n_4 = IJ(K - 1)$	$SS4 = \sum_i \sum_j \sum_k (Y_{ijk} - \bar{Y}_{ij*})^2$	S_4^2	$\theta_4 = \sigma_E^2$
Total	$IJK - 1$	$SST = \sum_i \sum_j \sum_k (Y_{ijk} - \bar{Y}_{***})^2$		

where A_i is the contribution of the ith operator, B_j is the contribution of the jth machine, $(AB)_{ij}$ is the operator-machine interaction, and E_{ijk} is the contribution of all other noncontrollable factors. In this example $I = 3$, $J = 4$, and $K = 2$. It is further assumed that A_i, B_j, $(AB)_{ij}$ and E_{ijk} are jointly independent normal random variables with means of zero and variances σ_A^2, σ_B^2, σ_{AB}^2, and σ_E^2, respectively. The ANOVA table for this model is shown in Table 2.2.2.

It is important to notice that each expected mean square (EMS) is a linear function of the variance components σ_A^2, σ_B^2, σ_{AB}^2, and σ_E^2. Setting confidence intervals on these variance components is accomplished by setting confidence intervals on linear combinations of the expected mean squares. For example, to obtain a confidence interval on σ_A^2, a confidence interval is obtained on the linear combination $(\theta_1 - \theta_3)/(KJ)$. The two-factor crossed random model shown in equation (2.2.2) is presented in Chapter 6.

As illustrated in these two examples, there are many parameters of interest in a variance component model. The next section discusses how one can make statistical inferences for these parameters.

2.3　POINT ESTIMATION, CONFIDENCE INTERVALS, AND TESTS OF HYPOTHESES

Three main procedures for making statistical inferences are point estimation, interval estimation, and hypothesis testing. Each of these concepts is now discussed.

2.3.1　Point Estimation

Point estimation concerns the selection of a function of the sample values that will "best" represent the parameter of interest. "Best" means that in some sense the estimate is close to the value of the unknown parameter. However, since the sample values are values of random variables, the estimate may be "close" to the parameter for some samples and "not close" to the parameter for other samples. In this book the word "best" means the estimate is (1) unbiased, and (2) has minimum variance in the class of unbiased estimators. We denote these as minimum variance unbiased (MVU) estimators. When MVU estimators for parameters exist, we report these formulas. However, they do not exist in a majority of cases and as pointed out by Searle (1971, p. 455), the relevance of unbiasedness is questionable in unbalanced random and mixed models. For these reasons, we either omit discussion of point estimation or provide selected estimators that satisfy other reasonable criteria for "bestness."

The statistical literature contains much research on point estimation of variance components. Readers who are primarily interested in point estimation should refer to the books by Searle, Casella, and McCulloch (1992), Rao and Kleffe (1988) and bibliographies provided by Sahai (1979), Sahai, Khuri, and Kapadia (1985), and Khuri and Sahai (1985). The primary focus of this book is interval estimation. We consider interval estimation to be more informative than point estimation because in most studies, it is generally not enough to obtain a single value for the parameter under investigation. For example, in determining the average tensile strength of wire, it may not be necessary to know the exact average tensile strength. Rather, a value within a half pound of the true average tensile strength will most likely be adequate for decision making. It is generally desirable, however, to have some confidence that specified limits contain the unknown parameter value. A point estimate provides no information concerning either the confidence or the limits. A confidence interval provides this information.

2.3.2   Confidence Intervals

Let θ represent a parameter of interest. A confidence interval is a random interval whose endpoints L and U, where $L \leq U$, are functions of the sample values such that $P[L \leq \theta \leq U] = 1 - 2\alpha$. The term $1 - 2\alpha$ is the confidence coefficient and is selected prior to data collection. We use the notation $1 - 2\alpha$ to denote the confidence coefficient of a two-sided interval and $1 - \alpha$ to denote the confidence coefficient of the one-sided intervals discussed later in this section. Typical values for the confidence coefficient are .90, .95, and .99. A $1 - 2\alpha$ confidence interval is represented as $[L;U]$ where U is the upper bound and L the lower bound. A confidence interval $[L;U]$ that satisfies $P[L \leq \theta \leq U] = 1 - 2\alpha$ is called an exact two-sided $1 - 2\alpha$ confidence interval. Often exact $1 - 2\alpha$ confidence intervals do not exist and $P[L \leq \theta \leq U]$ is only approximately equal to $1 - 2\alpha$. These intervals are referred to as approximate intervals. An approximate interval is conservative if $P[L \leq \theta \leq U] > 1 - 2\alpha$ and liberal if $P[L \leq \theta \leq U] < 1 - 2\alpha$. As a general rule, conservative intervals are preferred when only approximate intervals are available. However, if it is known that the actual confidence coefficient of a liberal interval is not much below $1 - 2\alpha$, the liberal interval can be recommended.

As discussed in Section 2.3.1, point estimators have been derived for variance components in a variety of statistical designs. Using these results, confidence intervals based on large sample normal theory can be formed. However, in most cases where sample sizes are small, these intervals cannot be recommended. The intervals presented in this book are not based on large sample normal theory. In contrast to the large sample intervals, the methods presented in this book provide "good" intervals for any sample size. We define a "good" confidence interval to be one that has a confidence coefficient equal or close to a specified value and thus provides useful information about the parameter of interest.

There are several criteria for defining the usefulness of a $1 - 2\alpha$ confidence interval for θ. One criterion is that of uniformly most accurate. This property means that the interval has a smaller probability of containing values not equal to θ than any other confidence interval with confidence coefficient $1 - 2\alpha$. Another desirable property is that of unbiasedness. A $1 - 2\alpha$ unbiased confidence interval is one where the probability of covering any value not equal to θ is less than or equal to $1 - 2\alpha$. A confidence interval that is *uniformly most accurate* (UMA) within the class of unbiased confidence intervals is referred to as the *uniformly most accurate unbiased* (UMAU) confidence interval. Finally, one may wish to consider only the length of an interval as an indicator of its information content. In this case, one might select an interval that is shorter or has shorter expected length than any other $1 - 2\alpha$ interval. All of these criteria have merit and might be preferred in any particular situation. However, except in the simplest cases, confidence intervals for variance component models that satisfy any of these criteria do not exist. In these cases, we present methods that give confidence intervals with a confidence coefficient close to $1 - 2\alpha$ and provide interval widths that are relatively short. For further discussion concerning properties of confidence intervals, the reader can consult Section 2.9 of Graybill (1976) and the references listed therein.

Discussion to this point has concerned confidence intervals where both L and U are random variables. In many situations, however, an investigator is interested in an interval where only one end point is random. An upper $1 - \alpha$ confidence interval on θ satisfies the equation $P[L \le \theta < \infty) = 1 - \alpha$ and a lower $1 - \alpha$ confidence interval on θ satisfies the equation $P(-\infty < \theta \le U] = 1 - \alpha$. The random variables L and U are the lower and upper bounds, respectively. In many books, the upper and lower confidence intervals are referred to as one-sided intervals. We adopt this terminology and in addition, refer to intervals of the form $[L;U]$ as "two-sided" intervals. The one-sided and two-sided intervals are related in the following manner. Let $[L;\infty)$ be an upper $1 - \alpha_1$ confidence interval on θ and $(-\infty;U]$ be a lower $1 - \alpha_2$ confidence interval on θ. If $P[U \ge L] = 1$, then the interval $[L;U]$ forms a two-sided $1 - \alpha_1 - \alpha_2$ interval on θ. In the formulas presented in this book we select $\alpha_1 = \alpha_2 = \alpha$.

2.3.3 Hypothesis Testing

Sometimes an investigator has interest in a particular value of a parameter and wants to use a statistical test to determine whether the data are consistent with this value. For a simple illustration, consider Example 2.2.1 and assume it is thought that there is little or no variation in the nitrogen content among trees in the orchard. In order to validate this assumption, a test of size α can be used to test the null hypothesis $H_o : \sigma_A^2 = 0$ against the alternative hypothesis $H_a : \sigma_A^2 > 0$. This test is performed by computing an F-statistic, $F_c = S_1^2/S_2^2$, and comparing it with a tabled F-value. We assume the reader is acquainted with these techniques. Selected tests of hypotheses are discussed for each model in this book.

2.3.4 Differences Between Confidence Intervals and Hypothesis Tests

Confidence intervals and tests of hypotheses are procedures for making statistical inferences that attach measures of uncertainty to the inferences. It is almost always the case, however, that confidence intervals are "uniformly more informative" than hypothesis tests for making decisions based on parametric values. Thus, hypothesis tests are seldom needed if confidence intervals are available. To illustrate this concept, consider the quality control problem in Example 2.2.2. In this example the quality control manager is interested in the proportion of the total variation in screen lengths that is due to the operators. This proportion is represented as $\rho_A = \sigma_A^2/(\sigma_A^2 + \sigma_B^2 + \sigma_{AB}^2 + \sigma_E^2)$. Machine operators are presently required to take a 10 hour training course before they operate the screen-making machines. The training program is very expensive and the manager would like to decrease the training hours if this can be done without affecting product variability. On the other hand, if the present level of training is not enough, the number of training hours must be increased.

If operators are properly trained, the manager expects a value of ρ_A close to .4. After the data from the experiment described in Example 2.2.2 have been collected, the manager constructs a two-sided confidence interval on ρ_A. Consider three possible resulting intervals: [.41; .44], [.80; .99], and [.41; .99]. All three of these intervals fail to contain the hypothesized value of .4 and a test of $H_o : \rho_A = .4$ against $H_a : \rho_A \neq .4$ would result in rejection of H_o. But the conclusion that results from each interval is different. In the first case, the manager would undoubtedly conclude that for practical purposes, the variation accounted for by the operators is close enough to .4 that increased training would not be worth the additional cost. In the second case, the evidence suggests that the operators are not well trained and the training program must be lengthened. In the final situation, the manager may decide that the confidence interval is too wide to be conclusive and that additional information is required before a decision can be made.

Now consider a situation where the confidence interval includes .4 and $H_o : \rho_A = .4$ would not be rejected. For example, two such intervals are [.35; .45] and [.1; .9]. In the first case the manager would undoubtedly conclude that the present training program is adequate whereas in the second case it might be decided that the investigation was inconclusive.

This example illustrates how confidence intervals afford the decision maker the opportunity to consider ''practical'' significance in addition to ''statistical'' significance. In addition, confidence intervals can be used to test hypotheses, whereas the results of a hypothesis test cannot be used to directly answer the questions posed in interval estimation. It is for these reasons that we argue confidence intervals are the most informative summary results of statistical inference and should be used whenever possible. Additional arguments for preferring confidence intervals to hypothesis tests are presented by Wonnacott (1987). The major objective of the book, therefore, is to present methods that provide "good" confidence intervals for making decisions. Although confidence intervals are the most informative single piece of information, many researchers also like to report point estimates and results of hypothesis tests. For this reason, we have included selected results for point estimation and hypothesis testing in the book.

2.4 SIMULTANEOUS CONFIDENCE INTERVALS

A confidence interval $[L;U]$ that satisfies $P[L \leq \theta \leq U] = 1 - 2\alpha$ is sometimes referred to as a one-at-a-time confidence interval. This type of interval is in contrast to the simultaneous confidence intervals discussed in this section. Simultaneous confidence intervals are a collection of confidence intervals for different functions of parameters based on the same sample. To illustrate the difference between one-at-a-time and simultaneous intervals, consider an investigator who desires two-sided confidence intervals on the parameters σ_A^2 and σ_B^2 in the two-factor crossed model of Example 2.2.2. The one-at-a-time confidence statements are

$$P[L_A \leq \sigma_A^2 \leq U_A] = 1 - 2\alpha$$
$$P[L_B \leq \sigma_B^2 \leq U_B] = 1 - 2\alpha \qquad (2.4.1)$$

When looked at individually, the probability is $1 - 2\alpha$ that each confidence statement in (2.4.1) is correct. If one wants to express the confidence that both intervals will simultaneously include the desired parameters, a probability statement of the form

$$P[L_A^* \leq \sigma_A^2 \leq U_A^* \quad \text{and} \quad L_B^* \leq \sigma_B^2 \leq U_B^*] = 1 - 2\alpha \quad (2.4.2)$$

is required. The intervals $[L_A^*; U_A^*]$ and $[L_B^*; U_B^*]$ are called simultaneous confidence intervals.

In any investigation there is general interest in making statistical inference about many parameters. For example, in Example 2.2.2 one would undoubtedly want confidence intervals on each of the variance components, σ_A^2, σ_B^2, σ_{AB}^2, and σ_E^2. In addition, interest may focus on the total variability, $\gamma = \sigma_A^2 + \sigma_B^2 + \sigma_{AB}^2 + \sigma_E^2$, or on the proportions of variability σ_A^2/γ, and σ_B^2/γ. One therefore must determine whether one-at-a-time or simultaneous confidence intervals are desired. We suggest the following guideline for making this decision. If an investigator is interested in making a decision that requires knowledge of several functions of parameter values simultaneously, and if the desired probability of a correct decision is $1 - 2\alpha$, then simultaneous confidence intervals should be used. In all other situations, one-at-a-time confidence intervals are recommended. An example is provided to illustrate situations in which each type of interval might be desired.

Suppose the manufacturer of a very sensitive instrument wants 99% of the instruments produced to operate within specified limits. To do this, it is necessary for each of three component parts of the instrument to operate within specified limits. The manufacturer samples the daily production of the parts and sets confidence limits on the variances of each part to determine if these variances are in compliance. In this case, 99% simultaneous confidence intervals are used for the set of three variances so that if each of the three confidence bounds is within its respective limit, the instrument will operate within the specified limits with probability 99%. In contrast, if one is interested in examining the variances of the three component parts separately and no single part or decision depends on the simultaneous performance of the three parts, then one-at-a-time confidence intervals should be used.

As might be expected, simultaneous confidence intervals are wider than one-at-a-time confidence intervals. Also, the procedures for computing the two types of intervals generally are different and require different tabled values. However, one special case in which simultaneous confidence intervals are constructed from one-at-a-time intervals deserves mention. If one-at-a-time confidence intervals are available for a set of parameters, then the Bonferroni inequality can be used to obtain simultaneous confidence intervals. Suppose Q one-at-a-time confidence statements on $\theta_1, \theta_2, \ldots, \theta_Q$ are

$$P\left[L_1 \leq \theta_1 \leq U_1\right] = 1 - 2\alpha_1$$
$$P\left[L_2 \leq \theta_2 \leq U_2\right] = 1 - 2\alpha_2$$
$$\vdots$$
$$P\left[L_Q \leq \theta_Q \leq U_Q\right] = 1 - 2\alpha_Q \tag{2.4.3}$$

The Bonferroni inequality (see, e.g., Miller, 1981) provides the following simultaneous confidence statement.

$$P[L_1 \le \theta_1 \le U_1, \quad \text{and} \quad L_2 \le \theta_2 \le U_2, \quad \text{and} \quad ..., \quad \text{and} \quad L_Q \le \theta_Q \le U_Q]$$
$$\ge 1 - 2\alpha_1 - 2\alpha_2 - \cdots - 2\alpha_Q. \tag{2.4.4}$$

Thus, the set of intervals $[L_1;U_1]$, $[L_2;U_2]$, ..., $[L_Q;U_Q]$ provides a conservative confidence region with confidence coefficient at least as great as $1 - 2\alpha_1 - 2\alpha_2 - \cdots - 2\alpha_Q$. This simultaneous set of intervals can be very conservative for $Q > 2$.

To illustrate (2.4.4), let $Q = 2$ and suppose $P[L_1 \le \theta_1 \le U_1] = 1 - 2\alpha_1 = .95$, and $P[L_2 \le \theta_2 \le U_2] = 1 - 2\alpha_2 = .90$. Thus, $\alpha_1 = .025$, $\alpha_2 = .05$, $1 - 2\alpha_1 - 2\alpha_2 = .85$, and we have the statement $P[L_1 \le \theta_1 \le U_1 \text{ and } L_2 \le \theta_2 \le U_2] \ge .85$. In words, the probability is greater than or equal to 85% that simultaneously the intervals $[L_1; U_1]$ and $[L_2; U_2]$ will contain θ_1 and θ_2, respectively.

The confidence intervals presented in this book are one-at-a-time intervals. Khuri (1981) developed an optimization technique that provides simultaneous confidence intervals on all continuous functions of variance components in a balanced random model. The technique also can be applied to some unbalanced random models. However, shorter simultaneous intervals can often be obtained by using the Bonferroni inequality to combine the one-at-a-time intervals presented in this book. A more complete discussion of this point is presented in Section 3.5.

2.5 NOTATION AND USE OF TABLES

In this section we present the notation that is used in the book. Chapters 3–6 consider several random models. These models are defined in terms of a standard model formulation and an ANOVA table. Chapter 7 considers mixed models that involve both fixed and random effects.

The general random model formulation is

$$Y_{ij...p} = \mu + A_i + B_j + \cdots + E_{ij...p}$$
$$i = 1, ..., I; j = ..., J; ...; p = 1, ..., P \tag{2.5.1}$$

where μ is a constant, and A_i, B_j, ..., $E_{ij...p}$ are jointly independent normal random variables with means of zero and variances σ_A^2,

σ_B^2, ..., and σ_E^2, respectively. In designs that are not complete and balanced (unequal sample sizes) the subscripts on the random effects will be modified. This will be noted when these designs are encountered. Random effects are denoted with upper case letters. The fixed effects encountered in Chapter 7 are denoted with Greek letters. Interaction terms denote which effects are involved, e.g., $(AB)_{ij}$ indicates AB interaction and $(ABC)_{ijk}$ indicates ABC interaction. The variance component associated with any random effect is represented as σ^2 with an appropriate subscript. For example, the variance of A_i is σ_A^2, the variance of B_j is σ_B^2, and the variance of $(AB)_{ij}$ is σ_{AB}^2.

The general ANOVA table for (2.5.1) is shown in Table 2.5.1. Table 2.5.1 consists of Q sources of variation. For example, $Q=2$ in Table 2.2.1 and $Q=4$ in Table 2.2.2. The degrees of freedom are denoted as n_q, the sums of squares as SSq, the mean squares as S_q^2, and the expected mean squares as θ_q for $q=1$, ..., Q. The definitions for the sums of squares are given in terms of the original observations $Y_{ijk\cdots p}$. Graybill and Hultquist (1961) have shown that in balanced complete random models these sums of squares can be obtained by treating the random effects as if they are fixed effects. When as asterisk (*) replaces a subscript, this means the observations have been summed over that subscript. For example, $Y_{i*k} = \Sigma_j^J Y_{ijk}$ and $Y_{*j*} = \Sigma_i^I \Sigma_k^K Y_{ijk}$. An overbar denotes a mean when summed over the subscript replaced with an asterisk. Thus, $\bar{Y}_{i*k} = \Sigma_j^J Y_{ijk} / J$ and $\bar{Y}_{*j*} = \Sigma_i^I \Sigma_k^K Y_{ijk} / (IK)$.

Before using any formula in this book, the reader should refer to the ANOVA table corresponding to the appropriate model. This is recommended because the definition assigned to any particular symbol may vary from model to model. For example, the definitions for n_2 and

Table 2.5.1 General ANOVA for a Variance Component Model

SV	DF	SS	MS	EMS
Factor A	n_1	$SS1$	S_1^2	θ_1
Factor B	n_2	$SS2$	S_2^2	θ_2
⋮	⋮	⋮	⋮	⋮
Error	n_Q	SSQ	S_Q^2	θ_Q

SS2 in Table 2.2.1 are different than the definitions of n_2 and *SS2* in Table 2.2.2.

As noted earlier, the expected mean squares are linear combinations of the variance components. To set confidence intervals on the variance components, one sets confidence intervals on appropriate linear combinations of the expected mean squares. The functions of the variance components that are of interest are sometimes complicated functions of the θ_q. Consider, for example, confidence intervals on ratios of linear combinations of variance components. For the one-factor design shown in Table 2.2.1, $\sigma_A^2/(\sigma_A^2 + \sigma_E^2) = (\theta_1 - \theta_2)/[\theta_1 + (J - 1)\theta_2]$ and for the two-factor design in Table 2.2.2, $\sigma_A^2/(\sigma_A^2 + \sigma_B^2 + \sigma_{AB}^2 + \sigma_E^2) = I(\theta_1 - \theta_3)/[I\theta_1 + J\theta_2 + (IJ - I - J)\theta_3 + (KIJ - IJ)\theta_4]$.

For balanced and complete designs, the assumption of joint independence among the normal random components in (2.5.1) implies the $n_q S_q^2/\theta_q$ terms in Table 2.5.1 are independent central chi-squared random variables with n_q degrees of freedom. Readers interested in a theoretical development of this result can consult Chapter 15 of Graybill (1976). In situations where unbalancedness exists, this property is lost for some of the $n_q S_q^2/\theta_q$, although the chi-squared distribution is often used for approximation. These distributional results require repeated use of tabled chi-squared and F-values for computing confidence intervals on functions of the variance components. Because chi-squared and F-values are related by the equation $\chi_{\alpha:\nu}^2 = \nu F_{\alpha:\nu,\infty}$ where ν represents the degrees of freedom associated with the chi-squared random variable, these values can all be found in an F-table.

Appendix A provides F-tables that can be used to compute the confidence intervals presented in the book. Use of these tables is illustrated by considering the problem of constructing a confidence interval on θ_1/θ_2 where $n_1 S_1^2/\theta_1$ and $n_2 S_2^2/\theta_2$ are independent chi-squared random variables with n_1 and n_2 degrees of freedom, respectively. Given these distributional assumptions,

$$P\left[\frac{S_1^2\theta_2}{S_2^2\theta_1} \leq F_{\alpha_1:n_1,n_2}\right] = 1 - \alpha_1 \qquad (2.5.2)$$

where $F_{\alpha_1:n_1,n_2}$ represents the F-value with n_1 and n_2 degrees of freedom that has α_1 area to the right. Statement (2.5.2) can be reorganized as

$$P\left[\frac{S_1^2}{S_2^2\,F_{\alpha_1:n_1,n_2}} \le \frac{\theta_1}{\theta_2}\right] = 1 - \alpha_1 \qquad (2.5.3)$$

The one-sided upper $1 - \alpha_1$ confidence interval on θ_1/θ_2 is therefore

$$\left[\frac{S_1^2}{S_2^2\,F_{\alpha_1:n_1,n_2}}\,;\,\infty\right) \qquad (2.5.4)$$

where $S_1^2/(S_2^2 F_{\alpha_1:n_1,n_2})$ is the lower bound on θ_1/θ_2. In a similar fashion, the one-sided lower $1 - \alpha_2$ confidence interval on θ_1/θ_2 is

$$\left[0;\,\frac{S_1^2}{S_2^2\,F_{1-\alpha_2:n_1,n_2}}\right] \qquad (2.5.5)$$

where $S_1^2\,/\,(S_2^2\,F_{1-\alpha_2:n_1,n_2})$ is the upper confidence bound on θ_1/θ_2.

Appendix A contains the F-values used in (2.5.2)–(2.5.5). These tables are used in the following manner. At the intersection of the row labeled r and the column labeled c in these tables are two F-values. These F-values are denoted $F_{1-\alpha:r,c}$ and $F_{\alpha:r,c}$ where $1 - 2\alpha$ denotes the two-sided confidence coefficient, r the row label, and c the column label. To illustrate, let $1 - 2\alpha = .80$ and use Table A.1 to find $F_{.90:3,5} = .1884$ and $F_{.10:3,5} = 3.6195$. In many of the examples considered in the book, the required degrees of freedom are not shown in Appendix A. In these situations, F-values are obtained using FINV of the Statistical Analysis System (SAS©).

A two-sided $1 - 2\alpha$ confidence interval on θ_1/θ_2 is

$$\left[\frac{S_1^2}{S_2^2\,F_{\alpha_1:n_1,n_2}}\,;\,\frac{S_1^2}{S_2^2\,F_{1-\alpha_2:n_1,n_2}}\right] \qquad (2.5.6)$$

where $\alpha_1 + \alpha_2 = 2\alpha$. There are an infinite number of values for α_1 and α_2 that satisfy $\alpha_1 + \alpha_2 = 2\alpha$, and hence some criterion for selecting appropriate values is needed. Possible criteria for selecting these values previously have been discussed in Section 2.3.2. Most authors and investigators use one of three methods. These methods are (1) equal-tailed ($\alpha_1 = \alpha_2 = \alpha$), (2) shortest, and (3) uniformly most accurate unbiased. All tables in Appendix A have $\alpha_1 = \alpha_2 = \alpha$. Tables of F-values that provide shortest intervals on θ_1/θ_2 in (2.5.6) have been reported by Levy and Narula (1974), Guenther (1977), and Lin, Richards, Long, Myers, and Taylor (1983). Conditions for deriving the F-values that provide the UMAU intervals on θ_1/θ_2 in (2.5.6) have been reported by Ramachandran (1958) and Guenther (1977).

For the approximate confidence intervals presented in this book, F-values that provide either shortest of UMAU confidence intervals are not known. In our research we have examined the performance of these intervals using equal-tailed F-values and the F-values that provide the shortest and UMAU intervals on θ_1/θ_2 in (2.5.6). Based on our work, we recommend using equal-tailed F-values in two-sided intervals for several reasons. First, when using equal-tailed F-values, an investigator will know the probability of missing the parameter on both the high and low sides. This is a decided advantage when these probabilities must be controlled. In fact, if the consequences associated with missing the parameter are different for the high and low sides, a combination of one-sided intervals with appropriate F-values is recommended. In contrast, probabilities of missing the parameter on either the high or the low side are not easily determined when F-values other than equal-tailed F-values are used in computations. A second disadvantage when using the shortest F-values associated with (2.5.6) is that they force the lower bound to zero and provide a two-sided interval that is conceptually more consistent with the notion of a one-sided lower interval.

For convenience, we present confidence intervals in the form of two-sided intervals. However, one-sided confidence intervals can always be obtained from these formulas. To illustrate, let $[L;U]$ represent a two-sided $1 - 2\alpha$ confidence interval based on the equal-tailed F-values of Appendix A. An upper $1 - \alpha$ confidence interval on the parameter is therefore $[L; \infty)$ and a lower $1 - \alpha$ confidence interval on the parameter is $(-\infty;U]$. In some cases these intervals are modified based on the parameter space. For example, since variance components cannot be negative, a lower interval on a variance component is $[0;U]$.

2.6 CONFIDENCE INTERVALS ON THE VARIANCE OF A NORMAL POPULATION

Consider a random sample $Y_1, Y_2, \ldots, Y_n$ selected from a normal population with unknown mean μ and unknown variance σ^2. Define the sample mean $\bar{Y} = \Sigma_i^n Y_i/n$ and sample variance $S^2 = \Sigma_i^n (Y_i - \bar{Y})^2/(n-1)$. Since $(n-1)S^2/\sigma^2$ has a chi-squared distribution with $n-1$ degrees of freedom, an exact two-sided $1 - 2\alpha$ confidence interval on σ^2 is

$$\left[\frac{S^2}{F_{\alpha:n-1,\infty}} \; ; \; \frac{S^2}{F_{1-\alpha:n-1,\infty}} \right] \qquad (2.6.1)$$

Klein (1990) examined the effect of n on the properties of the length of (2.6.1) and notes that when n is small, (2.6.1) has a large expected width with respect to σ^2 and a large probability of false coverage. As n increases, there is dramatic improvement until n attains a modest level (approximately 40). There is not much appreciable gain as n is increased beyond this point.

Tables of F-values that provide the shortest two-sided interval on σ^2 in (2.6.1) have been reported in the form of chi-squared values by Tate and Klett (1959). Tables of F-values that provide the UMAU confidence interval on σ^2 in (2.6.1) have been reported by Ramachandran (1958), Tate and Klett (1959), Lindley, East, and Hamilton (1960), Pachares (1961), and John (1973). Guenther (1972) provides additional information concerning these computations.

Schulz (1976) compares three methods for constructing intervals on σ^2 when the underlying distribution is not normal. Nagata (1989) compares several procedures for constructing confidence intervals on σ^2 based on S^2 and S_o^2 where S_o^2 is independent of S^2 and $n_o S_o^2/\sigma^2$ has a non-central chi-squared distribution.

In many applications an interval on σ is more meaningful than an interval on σ^2. Exact confidence intervals on σ are obtained by taking the square roots of the bounds in interval (2.6.1). A numerical example is used to illustrate these calculations.

Example 2.6.1   A machine is used to fill bottles with vegetable oil. A random sample of six bottles is selected from the filled bottles and the oil is weighed in each bottle. The weights (in ounces) are as follows:

$$15.66, \ 15.66, \ 15.70, \ 15.70, \ 15.68, \ 15.70$$

The calculated sample variance is $S^2 = .0003867$. To construct an upper 95% confidence interval on the population variance of the bottle weights, we use the lower bound of (2.6.1) and Table A.2 where $F_{.05:5,\infty} = 2.2141$. The calculated interval is $[.0003867/2.2141; \ \infty) = [.000175; \ \infty)$. The calculated lower 95% confidence interval using the upper bound of (2.6.1) and $F_{.95:5,\infty} = .2291$ is $[0; .0003867/.2291] = [0; .00169]$. A two-sided 90% interval on the population variance is therefore $[.000175; \ .00169]$. The corresponding 90% confidence interval on the population standard deviation is $[.013; \ .041]$.

2.7 CONFIDENCE INTERVALS ON THE RATIO OF VARIANCES FROM TWO NORMAL POPULATIONS

Consider two independent random samples $X_{11}, X_{12}, \ldots, X_{1m}$ and $X_{21}, X_{22}, \ldots, X_{2n}$ selected from normal populations with means μ_1 and μ_2, and variances σ_1^2 and σ_2^2, respectively. Defining $\bar{X}_1 = \Sigma_i^m X_{1i}/m$, $\bar{X}_2 = \Sigma_j^n X_{2j}/n$, $S_1^2 = \Sigma_i^m (X_{1i} - \bar{X}_1)^2/(m-1)$, and $S_2^2 = \Sigma_j^n (X_{2j} - \bar{X}_2)^2/(n-1)$, it is known that $(S_1^2/\sigma_1^2)/(S_2^2/\sigma_2^2)$ has an F-distribution with $m-1$ and $n-1$ degrees of freedom. This is the same problem considered in Section 2.5 with $\theta_1 = \sigma_1^2$, $\theta_2 = \sigma_2^2$, $n_1 = m-1$, and $n_2 = n-1$. Thus, one-sided and two-sided confidence intervals on σ_1^2/σ_2^2 can be obtained from equations (2.5.4), (2.5.5), and (2.5.6). An example of the two population problem is presented.

Example 2.7.1   Continuing with the bottle data in Example 2.6.1, suppose a random sample of four bottles is selected from a second filling machine. The weights (in ounces) of these sampled bottles are

$$15.78, \ 15.80, \ 15.78, \ 15.79$$

Let S_1^2 denote the sample variance of the weights for the bottles filled on the first machine and S_2^2 denote the sample variance of the weights for the bottles filled on the second machine. Then $S_1^2 = .0003867$, $S_2^2 = .00009167$, and a two-sided 95% confidence interval on σ_1^2/σ_2^2 is obtained using (2.5.6) with $F_{.025:5,3} = 14.8848$ and $F_{.975:5,3} = .1288$ taken from Table A.3. The calculated interval is $[.284; 32.8]$. Since the value 1 is contained in the interval, the data are not sufficient to conclude there is a difference in the variability between the two filling machines.

The confidence intervals in (2.5.4)–(2.5.6) can be used to test $H_o : \sigma_1^2/\sigma_2^2 = c$ where c is a specified constant. In particular, if c is contained in the interval, the null hypothesis is not rejected. Intervals (2.5.4) and (2.5.5) provide tests for one-sided alternatives and (2.5.6) provides a test for a two-sided alternative. Balakrishnan and Ma (1990) review tests of $H_o : \sigma_1^2/\sigma_2^2 = 1$ when normality cannot be assumed.

Tests of hypotheses concerning equality of variances for several populations have been proposed by several authors. Kirk (1982, p. 77–79) and Milliken and Johnson (1984, p. 18–23) provide good reviews of these procedures. Ramachandran (1956) derives simultaneous confidence intervals on all ratios of the variances for several independent populations.

2.8 SUMMARY

This chapter has provided general results and notation that are used for the variance component models presented in the book. Before using any formulas, the reader is encouraged to refer to the appropriate model and ANOVA table. Although the tables found in Appendix A have been developed to facilitate location of F-values, the reader may wish to review the examples presented in this chapter before using the tables.

3

General Results for Balanced Designs

3.1 INTRODUCTION

This chapter presents general results that can be used for constructing confidence intervals on certain functions of variance components in balanced random and mixed designs. As mentioned in Chapter 2, functions of variance components are expressible as functions of expected mean squares. In the most general form, parameters of interest can be represented as either $\theta = \Sigma_q^Q c_q\theta_q$ or $\theta = \Sigma_q^Q c_q\theta_q / \Sigma_q^Q d_q\theta_q$ where c_q and d_q are selected constants, and the θ_q are expected mean squares for $q = 1, \ldots, Q$. For example, the ratio $\sigma_A^2/(\sigma_A^2 + \sigma_B^2 + \sigma_{AB}^2 + \sigma_E^2)$ for the two-factor design in Table 2.2.2 is written $I(\theta_1 - \theta_3)/[I\theta_1 + J\theta_2 + (IJ - I - J)\theta_3 + (KIJ - IJ)\theta_4]$. In terms of the general notation, $Q = 4$, $c_1 = I, c_2 = 0$, $c_3 = -I$, $c_4 = 0$, $d_1 = I$, $d_2 = J$, $d_3 = IJ - I - J$, and $d_4 = KIJ - IJ$.

For the balanced random normal probability models, the random variables $n_1S_1^2/\theta_1$, $n_2S_2^2/\theta_2$, $\ldots$, $n_QS_Q^2/\theta_Q$ in Table 2.5.1 represent a set of jointly independent chi-squared random variables with $n_1, n_2, \ldots, n_Q$ degrees of freedom, respectively. Exact or approximate confidence intervals have been developed for many functions of the θ_q under these distributional assumptions. These intervals are presented in this chapter and applied throughout the book for several experimental designs. Although there are many different variance component models, most parameters of interest can be represented in one of the forms presented in this chapter.

3.2 INTERVALS ON SUMS OF EXPECTED MEAN SQUARES

This section presents methods for constructing confidence intervals on $\gamma = \sum_q^Q c_q \theta_q$ where $c_q \geq 0$ and $n_q S_q^2/\theta_q$ for $q = 1, \ldots, Q$ are independently distributed chi-squared random variables with n_q degrees of freedom. The parameter γ is often used to represent the total variance of a response variable. For example, in the one-fold nested design of Table 2.2.1, $Q = 2$, $\theta_1 = \sigma_E^2 + J\sigma_A^2$, $\theta_2 = \sigma_E^2$, and the variance of $Y_{ij} = \gamma = \sigma_A^2 + \sigma_E^2$. Thus, $\gamma = c_1\theta_1 + c_2\theta_2$ where $c_1 = 1/J$ and $c_2 = 1 - 1/J$. The methods in this section all assume $c_q \geq 0$. If some $c_q < 0$, these methods may not work well and the procedures in Section 3.3 are recommended.

3.2.1 Intervals on γ when $Q = 1$

Exact intervals on θ_q are computed using the results of Section 2.6. Since $n_q S_q^2/\theta_q$ has a chi-squared distribution with n_q degrees of freedom, an exact two-sided $1 - 2\alpha$ confidence interval on θ_q is

$$\left[\frac{S_q^2}{F_{\alpha:n_q,\infty}} ; \frac{S_q^2}{F_{1-\alpha:n_q,\infty}} \right] \tag{3.2.1}$$

We now consider methods for constructing intervals on γ when $Q > 1$.

3.2.2 Satterthwaite's Procedure

A popular approach for constructing approximate intervals on γ was first proposed by Smith (1936) and later by Satterthwaite (1941, 1946). This approach is based on a chi-squared approximation of the MVU estimator for $\gamma, \hat{\gamma} = \sum_q^Q c_q S_q^2$. In particular, one determines the value of m that equates the first two moments of $m\hat{\gamma}/\gamma$ to those of a chi-squared random variable having m degrees of freedom. After replacing θ_q with S_q^2 in the formula for m, an approximate two-sided $1 - 2\alpha$ confidence interval on γ is

$$\left[\frac{\hat{\gamma}}{F_{\alpha:m,\infty}} ; \frac{\hat{\gamma}}{F_{1-\alpha:m,\infty}} \right] \qquad (3.2.2)$$

where

$$m = \hat{\gamma}^2 \Big/ \sum_q^Q \frac{c_q^2 S_q^4}{n_q}$$

In practice, m is generally not an integer and one often uses the greatest integer less than or equal to m. The Satterthwaite approximation works well when the n_q values are all equal or all large. However, when differences among the n_q are great, the Satterthwaite approximation can produce unacceptably liberal confidence intervals. Lower intervals (upper bounds) based on this approximation are more liberal than upper intervals (lower bounds). The Satterthwaite method should not be used if some of the c_q are negative and some are positive.

3.2.3 Welch's Procedure

For large values of n_q, large sample normal theory can be used to construct approximate confidence intervals on γ. These intervals are based on the result that $Z = (\hat{\gamma} - \gamma)/\sqrt{\text{Var}(\hat{\gamma})}$ has a limiting normal distribution with mean zero and variance one as $\min(n_1, n_2, ..., n_Q)$ approaches infinity where $\text{Var}(\hat{\gamma}) = 2\sum_q^Q c_q^2\theta_q^2/n_q$. The large sample two-sided $1 - 2\alpha$ confidence interval on γ is

$$\left[\hat{\gamma} - N_\alpha \sqrt{\text{Var}(\hat{\gamma})}; \hat{\gamma} + N_\alpha\sqrt{\text{Var}(\hat{\gamma})} \right] \qquad (3.2.3)$$

where N_α is the upper α probability point of a standard normal probability distribution. In applications, one replaces θ_q with S_q^2 in $\text{Var}(\hat{\gamma})$. Although interval (3.2.3) performs well asymptotically, it cannot be recommended for small values of n_q. Interval (3.2.3) can be improved for small values of n_q, however, by performing suitable modifications.

One such modification was suggested by Welch (1956) who considered the $1 - \alpha$ upper confidence interval $[L,\infty)$ where $L = \hat{\gamma} - N_\alpha \sqrt{2\sum_q^Q c_q^2 S_q^4 / n_q}$. He noted that $P[\gamma > L] = 1 - \alpha +$ terms of the order $\max(n_q^{-1/2})$. Using a series approximation, Welch added terms to L and reduced the approximation error from order $\max(n_q^{-1/2})$ to order $\max(n_q^{-3/2})$. He further noted $\hat{\gamma}/\gamma = [1 - (\hat{\gamma} - \gamma)/\hat{\gamma}]^{-1}$, and used the binomial theorem to express the approximation as an interval on the ratio $\hat{\gamma}/\gamma$. In this form, the lower bound of an upper $1 - \alpha$ confidence interval on γ is

$$L = \hat{\gamma} C^{-1} \tag{3.2.4}$$

where

$$C = 1 - e + dN_\alpha(1 - 2e + 18f - 2g/d^2 - 23e^2/(2d^2))$$

$$+ N_\alpha^2(d^2 - 2e) + dN_\alpha^3(d^2 - 4e - 8e^2/d^2 + 10f)$$

$$d = (2t_{21})^{1/2}/\hat{\gamma} \quad e = 4t_{32}/(3d^2\hat{\gamma}^3) \quad f = t_{43}/(d\hat{\gamma})^4$$

$$g = t_{22}/\hat{\gamma}^2 \quad \text{and} \quad t_{ab} = \sum_q^Q c_q^a S_q^{2a} n_q^{-b}$$

This result also was derived using alternative approaches by Bartlett (1953) and Huitson (1955). Welch used (3.2.4) to illustrate that his approximation can be viewed as a refinement of the Satterthwaite approximation. Lower $1 - \alpha$ confidence intervals are obtained from (3.2.4) by replacing N_α with $N_{1-\alpha}$. A $1 - 2\alpha$ two-sided interval on γ is formed by the bounds of the one-sided intervals. As noted with the Satterthwaite procedure, Welch's method should not be used unless all the c_q values are positive.

3.2.4 Modified Large-Sample Procedures

Graybill and Wang (1980) proposed a different modification of (3.2.3) for constructing a two-sided interval on γ. This approach makes the interval on γ exact for certain special cases with the hope that it will be close to exact in any alternative situation. Graybill and Wang termed this approach the Modified Large-Sample (MLS) method. This method provides an exact interval on γ when any $Q - 1$ of the θ_q are zero or when any $Q - 1$ of the n_q approach infinity. The Graybill-Wang two-sided $1 - 2\alpha$ interval on γ is

$$\left[\hat{\gamma} - \sqrt{\sum_q^Q G_q^2 c_q^2 S_q^4}; \; \hat{\gamma} + \sqrt{\sum_q^Q H_q^2 c_q^2 S_q^4} \right] \qquad (3.2.5)$$

where

$$G_q = 1 - \frac{1}{F_{\alpha:n_q,\infty}} \qquad \text{and}$$

$$H_q = \frac{1}{F_{1-\alpha:n_q,\infty}} - 1$$

The confidence coefficient for any interval on γ depends on the parameters $\rho_q = c_q \theta_q / \gamma$ for $q = 1, \ldots, Q - 1$. Graybill and Wang selected several sets of $(n_1, n_2, \ldots, n_Q)$ and varied the values of ρ_q across all feasible ranges for $Q = 2$ and $Q = 3$. They used numerical integration and simulation to compute confidence coefficients of Satterthwaite, Welch, and Graybill-Wang two-sided intervals on γ. Tables 3.2.1 and 3.2.2 report some of their results for $Q = 2$.

The tables report ranges of the true confidence coefficients across all feasible values of ρ_q. The values were computed using numerical

Table 3.2.1 Ranges of 90% Confidence Coefficients for Two-Sided Intervals on $\gamma = \Sigma_q^2 c_q \theta_q$ Using Satterthwaite (S), Welch (W), and Graybill-Wang (GW) (Results Based on Numerical Integration)

n_1	n_2	S	W	GW
5	5	.886–.921	.893–.907	.900–.917
5	30	.855–.912	.879–.915	.900–.916
10	10	.892–.913	.896–.904	.900–.910
10	30	.882–.908	.894–.903	.900–.908
30	30	.897–.905	.899–.901	.900–.904

Table 3.2.2   Ranges of 95% Confidence Coefficients for Two-Sided Intervals on $\gamma = \Sigma_q^2 \, c_q \theta_q$ Using Satterthwaite (S), Welch (W), and Graybill-Wang (GW) (Results Based on Numerical Integration)

n_1	n_2	S	W	GW
5	5	.939–.963	.942–.952	.949–.960
5	30	.907–.958	.928–.964	.950–.960
10	10	.943–.958	.946–.953	.950–.956
10	30	.934–.956	.944–.952	.950–.955
30	30	.948–.953	.949–.951	.950–.952

integration. These results indicate that the Welch method is better than the Satterthwaite method because it provides confidence coefficients closer to the stated level. However, as the differences among the n_q increase, both methods result in intervals with confidence coefficients that are much less than the stated level. As shown in the tables, Satterthwaite 90% (95%) confidence intervals have actual confidence coefficients as low as 85.5% (90.7%). Welch 90% (95%) confidence intervals have actual confidence coefficients as low as 87.9% (92.8%). These minimum values occur in the design with $n_1 = 5$ and $n_2 = 30$. The Satterthwaite and Welch intervals can be even more liberal for values of $n_1 < 5$ with $n_2 - n_1 > 20$.

In contrast, the confidence coefficients associated with (3.2.5) in the tables are at least as great as the stated level in virtually every case. Unlike the Satterthwaite and Welch approximations, the Graybill-Wang approximation improves when one n_q is fixed and the other degrees of freedom increase. Graybill and Wang also computed the expected lengths of the confidence intervals and found that (3.2.5) compares very favorably with the two-sided Welch interval. In particular, (3.2.5) was much shorter than the Satterthwaite and Welch two-sided intervals when the shortest and UMAU F-values for θ_q were used in the formulas. The length of (3.2.5) also compared favorably with the Satterthwaite and Welch intervals when equal-tailed F-values were used.

Lu (1985) performed an extended study of two-sided intervals on γ for $Q = 3$ and $Q = 4$. Her results were consistent with those reported by Graybill and Wang. In addition, she developed a modification of the

Graybill-Wang intervals that produces confidence coefficients closer to the stated level. Although this modification provides slightly better intervals, it is not described here because it requires the use of an optimization program.

The performance of one-sided intervals on γ were studied by Ting, Burdick, Graybill, and Gui (1989) and Wang (1988b, 1991). One-sided $1 - \alpha$ Graybill-Wang intervals are formed from (3.2.5) by using the appropriate bound. The lower bound in (3.2.4) provides a Welch upper $1 - \alpha$ confidence interval, and a Welch lower $1 - \alpha$ interval is obtained from (3.2.4) by replacing N_α with $N_{1-\alpha}$. Ting, Burdick, Graybill, and Gui (1989) compared these one-sided intervals for $Q = 2, 3, 4$, and 5. Tables 3.2.3 and 3.2.4 report ranges of the true confidence coefficients of one-sided intervals across feasible values of ρ_q for $Q = 2$ and $Q = 5$. Numerical integration was used to compute the values in Table 3.2.3 and simulation was used in Table 3.2.4. The simulated values were based on 10,000 iterations. This results in less than a 5% chance that the simulated value differs from the true value by more than .004.

Table 3.2.3 Ranges of 95% Confidence Coefficients for One-Sided Intervals on $\gamma = \Sigma_q^2 c_q \theta_q$ Using Satterthwaite (S), Welch (W), and Graybill-Wang (GW) (Results Based on Numerical Integration)

n_1	n_2	Lower intervals $[0; U]$		
		S	W	GW
5	5	.938–.968	.943–.957	.950–.985
5	30	.892–.956	.932–.965	.950–.979
10	10	.941–.960	.946–.955	.950–.978
10	30	.925–.955	.944–.953	.950–.973
30	30	.944–.954	.949–.951	.950–.967

n_1	n_2	Upper intervals $[L; \infty)$		
		S	W	GW
5	5	.950–.964	.950–.951	.928–.950
5	30	.950–.975	.947–.953	.931–.950
10	10	.950–.960	.950–.950	.931–.950
10	30	.950–.965	.949–.951	.933–.950
30	30	.950–.956	.950–.950	.936–.950

Table 3.2.4 Simulated Ranges of 95% Confidence Coefficients for One-Sided Intervals on $\gamma = \Sigma_q^5 c_q \theta_q$ Using Satterthwaite (S), Welch (W), and Graybill-Wang (GW)

n_1	n_2	n_3	n_4	n_5	S	W	GW
				Lower intervals [0; U]			
5	5	5	5	5	.923–.973	.930–.958	.948–.999
5	5	10	30	30	.866–.967	.925–.965	.946–.993
10	10	10	30	30	.907–.962	.937–.959	.949–.990
30	30	30	30	30	.931–.957	.946–.954	.946–.978
n_1	n_2	n_3	n_4	n_5	S	W	GW
				Upper intervals [L; ∞)			
5	5	5	5	5	.949–.975	.945–.954	.910–.951
5	5	10	30	30	.948–.985	.943–.965	.914–.953
10	10	10	30	30	.949–.976	.943–.956	.919–.953
30	30	30	30	30	.950–.965	.947–.955	.926–.954

These results indicate that the Welch lower bound in (3.2.4) provides almost exact upper intervals over a wide range of n_q and ρ_q. This approximation performs less well when differences among the degrees of freedom are large, although the confidence coefficient is always close to the stated level. As shown in the tables, the true confidence coefficients of Welch upper 95% intervals are never less than 94.3% nor greater than 96.5%. In contrast, the confidence coefficients of Graybill-Wang upper intervals are typically less than the stated level. The true confidence coefficients of Graybill-Wang upper 95% confidence intervals range from 91.0% to 95.4%. Thus, the Welch approximation is recommended for constructing upper intervals (lower bounds) on γ.

For lower intervals (upper bounds) on γ, the Welch interval cannot be recommended because it does not always maintain the stated confidence level. Although the lower Graybill-Wang interval can be extremely conservative in some cases, it generally maintains the stated level and is the recommended upper bound. Ting et al. (1989) proposed an alternative upper bound that is less conservative than the Graybill-Wang upper bound, although it can be slightly liberal for $Q = 2$.

Wang (1988b) compared the performance of Satterthwaite,

Welch, and Graybill-Wang one-sided intervals on γ for $Q = 2$ and obtained the same results as Ting et al. (1989). In addition, he developed an iterative method based on (3.2.5) to construct nearly exact upper and lower intervals on γ. Wang (1991) also proposed weighted averages of liberal and conservative confidence bounds.

As a final note, Burdick and Sielken (1978) considered the one-fold nested design and developed a procedure that provides exact one-sided and two-sided confidence intervals on γ when $Q = 2$. This method was improved by Seely (1980) and extended to any mixed model with two variance components and to some special cases where $Q > 2$. However, this exact method produces intervals that typically are much wider than the approximate intervals discussed in this section. Additionally, the method is not investigator invariant. This means that two investigators could use the same set of data and compute two different intervals. For these reasons, this exact method is not recommended.

This section has considered linear functions of expected mean squares with non-negative coefficients $(c_q \geq 0)$. In many situations, however, negative coefficients are necessary to define the parameter of interest. For example, in the one-fold nested design of Table 2.2.1, $\sigma_A^2 = (\theta_1 - \theta_2)/J$. The approximations discussed in this section will not generally work well with negative c_q values. In these situations, the methods presented in the next section are recommended.

3.3 CONFIDENCE INTERVALS ON LINEAR COMBINATIONS OF EXPECTED MEAN SQUARES WITH DIFFERENT SIGNS

The problem considered in this section concerns construction of a confidence interval on the parameter $\delta = \sum_{q=1}^{P} c_q\theta_q - \sum_{r=P+1}^{Q} c_r\theta_r$ where $c_q, c_r \geq 0$ and δ is unrestricted in sign. Examples of this parameter are found in Table 2.2.1 where $\sigma_A^2 = (\theta_1 - \theta_2)/J$, and in Table 2.2.2, where $\sigma_A^2 = (\theta_1 - \theta_3)/(KJ)$ and $\sigma_B^2 = (\theta_2 - \theta_3)/(KI)$. As these examples illustrate, individual variance components are formed as differences of expected mean squares. For this reason, methods for constructing confidence intervals on δ are needed.

In Section 3.3.1 we present results for the important special case where $Q = 2$ and $P = 1$, i.e., where $\delta = c_1\theta_1 - c_2\theta_2$. An extension of these results to the general case where $Q > 2$ is provided in Section 3.3.2.

3.3.1 Intervals on $\delta = c_1\theta_1 - c_2\theta_2$

We first consider the situation where an interval on $\delta = c_1\theta_1 - c_2\theta_2$ ($c_1, c_2 \geq 0$) is desired, and it is known that $\delta \geq 0$. For example, δ might represent an individual variance component. Since δ includes a negative coefficient on θ_2, the Satterthwaite and Welch procedures discussed in Section 3.2 are not recommended. In such situations these methods produce intervals that are often very liberal. Healy (1961) developed a procedure that provides an exact interval on δ, but it requires a randomization device. This has the disadvantage of producing different intervals for the same observed data set. In addition, the intervals may sometimes contain gaps.

Many authors have developed approximate intervals on δ and Boardman (1974) conducted a study to compare several of these methods. Using Monte Carlo simulation, Boardman compared the confidence coefficients and interval lengths for two-sided equal-tailed intervals. The results of the study indicate that two procedures provide confidence coefficients close to the stated level and can be recommended. One of these methods was developed independently by Tukey (1951) and Williams (1962), and the other method was developed independently by Moriguti (1954) and Bulmer (1957). Mostafa and Ahmad (1986) performed additional comparisons of the Tukey-Williams and Moriguti-Bulmer methods for various designs and error structures. They concluded that the two methods were similar across all the examined conditions. Wang (1990a) derived a lower bound on the confidence coefficient of the Tukey-Williams interval.

In the same year that Boardman's results were published, Howe (1974) suggested another method for constructing a one-sided $1 - \alpha$ confidence interval on δ. The method employs a Cornish-Fisher expansion that is modified to make it exact under certain conditions. The Howe interval possesses several desirable properties described by Bulmer (1957) and Scheffé (1959). Briefly, in addition to possessing an invariance property, the interval becomes exact as n_1 approaches infinity and n_2 is fixed, or as n_2 approaches infinity and n_1 is fixed. In addition, when n_1 and n_2 both approach infinity with n_1/n_2 fixed, the confidence coefficient approaches $1 - \alpha$. Wang (1990b) compared the intervals proposed by Tukey-Williams, Moriguti-Bulmer, and Howe. He concluded that all the intervals generally maintain the stated confidence coefficient and that all three methods can be recommended in practice.

In some situations, the sign of δ is unknown. For example if $I = J$ in Table 2.2.2, $(\theta_1 - \theta_2)/KJ = \sigma_A^2 - \sigma_B^2$ and it is not necessarily known whether σ_A^2 is greater than σ_B^2. In Section 4.2.5, we present a model formulation proposed by Smith and Murray (1984) where δ represents a covariance and can be either negative or positive. As a final example, Samaranayake and Bain (1988) consider a test of equality for means of two exponential distributions that can be formulated as a test for $\delta = 0$ where the sign of δ is unknown.

Lu, Graybill, and Burdick (1988) proved a theorem that can be used to generalize the methods developed for $\delta \geq 0$ to the case where the sign of δ is unknown. They illustrated this theorem by applying it to the method proposed by Howe. In an independent paper, Samaranayake and Bain (1988) suggested the same procedure and applied it to the method proposed by Tukey and Williams. Finally, Ting, Burdick, Graybill, Jeyaratnam, and Lu (1990) proposed an MLS-type interval that can be used when the sign of δ is unknown. They compared this interval to the generalized Howe interval and noted the performances of the two intervals were similar.

Based on these studies, it appears there is not much practical difference in the performances of the proposed methods. For this reason, we present only the intervals developed by Ting, Burdick, Graybill, Jeyaratnam, and Lu (1990). These intervals are similar in form to the Graybill-Wang intervals presented in Section 3.2, are simple to compute, and can be used without further modifications if the sign of δ is unknown. Additionally, it will be shown in Section 3.3.2 that these intervals can be generalized to the case where $Q > 2$. The lower bound on an upper $1 - \alpha$ confidence interval on δ is

$$L = \hat{\delta} - \sqrt{V_L} \qquad (3.3.1)$$

where

$$\hat{\delta} = c_1 S_1^2 - c_2 S_2^2 \qquad V_L = G_1^2 c_1^2 S_1^4 + H_2^2 c_2^2 S_2^4 + G_{12} c_1 c_2 S_1^2 S_2^2$$

$$G_1 = 1 - \frac{1}{F_{\alpha:n_1,\infty}}$$

$$H_2 = \frac{1}{F_{1-\alpha:n_2,\infty}} - 1 \quad \text{and}$$

$$G_{12} = \frac{(F_{\alpha:n_1,n_2} - 1)^2 - G_1^2 F_{\alpha:n_1,n_2}^2 - H_2^2}{F_{\alpha:n_1,n_2}}$$

The upper bound of the lower $1 - \alpha$ confidence interval on δ is

$$U = \hat{\delta} + \sqrt{V_U} \tag{3.3.2}$$

where

$$V_U = H_1^2 c_1^2 S_1^4 + G_2^2 c_2^2 S_2^4 + H_{12} c_1 c_2 S_1^2 S_2^2$$

$$H_1 = \frac{1}{F_{1-\alpha:n_1,\infty}} - 1$$

$$G_2 = 1 - \frac{1}{F_{\alpha:n_2,\infty}} \quad \text{and}$$

$$H_{12} = \frac{(1 - F_{1-\alpha:n_1,n_2})^2 - H_1^2 F_{1-\alpha:n_1,n_2}^2 - G_2^2}{F_{1-\alpha:n_1,n_2}}$$

If it is known that $\delta \geq 0$, any negative bounds are defined to be zero. Intervals (3.3.1) and (3.3.2) are exact when either (i) $c_1\theta_1 = 0$, or (ii) $c_2\theta_2 = 0$, or (iii) $c_1\theta_1 = c_2\theta_2$. Condition (iii) can be verified by noting $L > 0$ when $c_1 S_1^2/(c_2 S_2^2) > F_{\alpha:n_1,n_2}$ and $U > 0$ when $c_1 S_1^2/(c_2 S_2^2) > F_{1-\alpha:n_1,n_2}$.

3.3.2 Intervals on δ with $Q > 2$

The bounds in (3.3.1) and (3.3.2) were extended to the case where $Q > 2$ and $Q > P$ by Ting, Burdick, Graybill, Jeyaratnam, and Lu (1990). Quite simply, new variance estimators and cross-product terms are added to (3.3.1) and (3.3.2) as Q and P increase. The lower bound for an upper $1 - \alpha$ interval on δ is

$$L = \hat{\delta} - \sqrt{V_L} \qquad (3.3.3)$$

where

$$\hat{\delta} = \sum_{q=1}^{P} c_q S_q^2 - \sum_{r=P+1}^{Q} c_r S_r^2$$

$$V_L = \sum_{q=1}^{P} G_q^2 c_q^2 S_q^4 + \sum_{r=P+1}^{Q} H_r^2 c_r^2 S_r^4$$

$$+ \sum_{q=1}^{P} \sum_{r=P+1}^{Q} G_{qr} c_q c_r S_q^2 S_r^2 + \sum_{q=1}^{P-1} \sum_{t>q}^{P} G_{qt}^* c_q c_t S_q^2 S_t^2$$

$$G_q = 1 - \frac{1}{F_{\alpha:n_q, \infty}} \qquad (q = 1, \ldots, P)$$

$$H_r = \frac{1}{F_{1-\alpha:n_r, \infty}} - 1 \qquad (r = P + 1, \ldots, Q)$$

$$G_{qr} = \frac{(F_{\alpha:n_q, n_r} - 1)^2 - G_q^2 F_{\alpha:n_q, n_r}^2 - H_r^2}{F_{\alpha:n_q, n_r}} \qquad \text{and}$$

$$G_{qt}^* = \left[\left(1 - \frac{1}{F_{\alpha:n_q+n_t, \infty}} \right)^2 \frac{(n_q + n_t)^2}{n_q n_t} - \frac{G_q^2 n_q}{n_t} - \frac{G_t^2 n_t}{n_q} \right] \Big/ (P - 1)$$

$$(t = q + 1, \ldots, P)$$

If $P = 1$ then G_{qt}^* is defined to be zero. The upper bound for a lower $1 - \alpha$ interval on δ is

$$U = \hat{\delta} + \sqrt{V_U} \qquad (3.3.4)$$

where

$$V_U = \sum_{q=1}^{P} H_q^2 c_q^2 S_q^4 + \sum_{r=P+1}^{Q} G_r^2 c_r^2 S_r^4$$

$$+ \sum_{q=1}^{P} \sum_{r=P+1}^{Q} H_{qr} c_q c_r S_q^2 S_r^2 + \sum_{r=P+1}^{Q-1} \sum_{u>r}^{Q} H_{ru}^* c_r c_u S_r^2 S_u^2$$

$$H_q = \frac{1}{F_{1-\alpha:n_q,\infty}} - 1 \quad (q = 1, \ldots, P)$$

$$G_r = 1 - \frac{1}{F_{\alpha:n_r,\infty}} \quad (r = P + 1, \ldots, Q)$$

$$H_{qr} = \frac{(1 - F_{1-\alpha:n_q,n_r})^2 - H_q^2 F_{1-\alpha:n_q,n_r}^2 - G_r^2}{F_{1-\alpha:n_q,n_r}} \quad \text{and}$$

$$H_{ru}^* = \left[\left(1 - \frac{1}{F_{\alpha:n_r+n_u,\infty}}\right)^2 \frac{(n_r + n_u)^2}{n_r n_u} - \frac{G_r^2 n_r}{n_u} - \frac{G_u^2 n_u}{n_r} \right] \Big/ (Q - P - 1)$$

$$(u = r + 1, \ldots, Q)$$

If $Q = P + 1$ then H_{ru}^* is defined to be zero. When $Q = 2$ and $P = 1$, (3.3.3) is equivalent to (3.3.1) and (3.3.4) is equivalent to (3.3.2). When $Q = P > 2$, one should refer to the recommendations given in Section 3.2.

Ting, Burdick, Graybill, Jeyaratnam, and Lu (1990) used computer simulation to demonstrate that (3.3.3) and (3.3.4) provide confidence coefficients close to the stated levels over a wide range of conditions. A partial set of results is shown in Table 3.3.1.

The confidence coefficients of (3.3.3) and (3.3.4) are functions of $\rho_i = \theta_i / \sum_{i=1}^{Q} \theta_i$ only and Table 3.3.1 reports ranges of simulated confidence coefficients across all feasible ranges of ρ_i. The simulations are based on 10,000 iterations. Using the normal approximation to the binomial, if the true confidence coefficient is 90% (95%), there is less than a 2.5% chance that a simulated confidence coefficient will be less than 89.4% (94.6%). As shown in Table 3.3.1, one-sided intervals are sometimes liberal, with upper intervals (lower bounds) generally more liberal than lower intervals (upper bounds). The simulated confidence coefficients of one-sided 95% intervals never drop below 92.8% for upper intervals or below 93.7% for lower intervals. Confidence coefficients for two-sided intervals based on (3.3.3) and (3.3.4) are very close to the stated level. In Table 3.3.1 the simulated 90% two-sided confidence coefficients never drop below 88.6%. The results in Table 3.3.1 include all linear combinations with $Q \leq 6$ for the reported values of $n_1, \ldots, n_6$. For example, the results of $\theta_1 - \theta_6$ with $n_1 = 5$,

Table 3.3.1   Simulated Ranges of 95% Confidence Coefficients of Lower and Upper Intervals and 90% Two-Sided Intervals on $\delta = \Sigma_{q=1}^{P} c_q \theta_q - \Sigma_{r=P+1}^{Q} c_r \theta_r$

+	+	+	+	+	−	Upper interval	Lower interval	Two-sided interval
n_1	n_2	n_3	n_4	n_5	n_6	$[L;\infty)$	$(-\infty;U]$	$[L;U]$
5	5	10	10	40	40	.928–.954	.946–.993	.895–.935
5	10	10	40	40	5	.929–.965	.944–.992	.895–.938
10	10	10	10	10	10	.932–.959	.945–.993	.893–.936

+	+	+	+	−	−	Upper interval	Lower interval	Two-sided interval
n_1	n_2	n_3	n_4	n_5	n_6	$[L;\infty)$	$(-\infty;U]$	$[L;U]$
2	4	8	32	1	16	.933–.993	.937–.997	.886–.974
2	8	16	32	1	4	.934–.994	.940–.996	.888–.977
3	3	3	27	9	9	.937–.980	.942–.998	.892–.962
5	5	40	40	10	10	.935–.978	.946–.991	.897–.950

+	+	+	−	−	−	Upper interval	Lower interval	Two-sided interval
n_1	n_2	n_3	n_4	n_5	n_6	$[L;\infty)$	$(-\infty;U]$	$[L;U]$
1	2	4	8	16	32	.943–.984	.943–.999	.895–.965
1	16	32	2	4	8	.938–.996	.940–.994	.886–.975
4	4	4	4	4	4	.943–.995	.939–.995	.894–.967
5	5	5	40	40	40	.943–.972	.943–.993	.892–.952
10	10	10	10	10	10	.942–.986	.942–.986	.897–.946

$n_6 = 40$ are included in the first line of the table. Thus, the table represents a "worst case" scenario. Intervals with smaller values of Q generally perform much better than the table might suggest.

Sections 3.2 and 3.3 presented results for constructing confidence intervals on linear combinations of expected mean squares. We now consider methods for constructing intervals on ratios of expected mean squares.

3.4 CONFIDENCE INTERVALS ON RATIOS OF EXPECTED MEAN SQUARES

The most general type of ratio of expected mean squares can be expressed as $\theta = \Sigma_q^Q c_q \theta_q / \Sigma_q^Q d_q \theta_q$ where c_q and d_q are unrestricted in sign. Although no general formulation for constructing an interval on θ has been proposed, results for several special cases have been developed. These methods are presented in this section.

3.4.1 Intervals on θ_1/θ_2

Exact intervals on θ_1/θ_2 are formed by noting $(S_1^2 \theta_2)/(S_2^2 \theta_1)$ has an F-distribution with n_1 and n_2 degrees of freedom. This result provides the exact two-sided $1 - 2\alpha$ confidence interval on θ_1/θ_2

$$\left[\frac{S_1^2}{S_2^2 F_{\alpha:n_1,n_2}} ; \frac{S_1^2}{S_2^2 F_{1-\alpha:n_1,n_2}} \right] \qquad (3.4.1)$$

3.4.2 Intervals on $\rho = \sum_{q=1}^P c_q \theta_q / \sum_{r=P+1}^Q c_r \theta_r$ with $c_q , c_r \geq 0$

A useful form for representing many ratios of interest is $\rho = \Sigma_{q=1}^P c_q \theta_q / \Sigma_{r=P+1}^Q c_r \theta_r$ where $c_q, c_r > 0$ and θ_q, θ_r are different linear combinations of variance components, $q = 1, ..., P$; $r = P + 1, ..., Q$. As an example, in Table 2.2.2 the ratio of total variance to error is $(\sigma_A^2 + \sigma_B^2 + \sigma_{AB}^2 + \sigma_E^2)/\sigma_E^2$. This ratio can be written in the general notation as $[I\theta_1 + J\theta_2 + (IJ - I - J)\theta_3]/(IJK\theta_4) + (K - 1)/K = \rho + (K - 1)/K$ where $P = 3$, $Q = 4$, $c_1 = I$, $c_2 = J$, $c_3 = IJ - I - J$, and $c_4 = IJK$.

Cochran (1951) proposed an approximate interval on ρ based on the Satterthwaite approximation discussed in Section 3.2.2. In particular, one equates the first two moments of $X_1 = m_1 \Sigma_{q=1}^P c_q S_q^2 / \Sigma_{q=1}^P c_q \theta_q$ and $X_2 = m_2 \Sigma_{r=P+1}^Q c_r S_r^2 / \Sigma_{r=P+1}^Q c_r \theta_r$ to those of chi-squared random variables with m_1 and m_2 degrees of freedom, respectively. Since X_1 and X_2 are independent, an approximate F-statistic with m_1 and m_2 degrees of freedom is $\sum_{q=1}^{P} c_q S_q^2 / (\sum_{r=P+1}^{Q} c_r S_r^2)$. Using this F-approximation, a two-sided $1 - 2\alpha$ interval on ρ is

$$\left[\frac{\hat{\rho}}{F_{\alpha:m_1,m_2}} \; ; \; \frac{\hat{\rho}}{F_{1-\alpha:m_1,m_2}} \right] \tag{3.4.2}$$

where

$$\hat{\rho} = \frac{\sum_{q=1}^{P} c_q S_q^2}{\sum_{r=P+1}^{Q} c_r S_r^2} \; ; \quad m_1 = \left(\sum_{q=1}^{P} c_q S_q^2 \right)^2 \Big/ \sum_{q=1}^{P} \frac{c_q^2 S_q^4}{n_q} \quad \text{and}$$

$$m_2 = \left(\sum_{r=P+1}^{Q} c_r S_r^2 \right)^2 \Big/ \sum_{r=P+1}^{Q} \frac{c_r^2 S_r^4}{n_r}$$

In practice, m_1 and m_2 are generally not integers, and one uses the greatest integer less than or equal to m_1 and m_2, respectively. As before, the Satterthwaite approximation works well when all degrees of freedom are of the same magnitude. However, interval (3.4.2) becomes liberal as the differences among either the n_q or the n_r increase.

Myers and Howe (1971) proposed an alternative to (3.4.2) that is a direct approximation of an F-statistic. They provide simulation results for the special case where $P = 2$ and $Q = 4$ that indicate the method provides shorter intervals on ρ than (3.4.2). However, Davenport (1975) illustrates that this approximation produces a more liberal interval than (3.4.2). Since (3.4.2) can be very liberal in some situations, the more liberal approximation of Myers and Howe is not recommended. Howe and Myers (1970) also proposed a refinement of (3.4.2) for the special case where $P = Q - 1$. Although this refinement improves the Satterthwaite approximation, it is still liberal in some cases and is more computationally intensive than (3.4.2).

Ting, Burdick, and Graybill (1991) used the bounds on δ shown in (3.3.3) and (3.3.4) to derive confidence intervals on ρ. To illustrate their approach, consider a lower bound on ρ, say L^*, such that $P[L^* \le \rho]$ is close to $1 - \alpha$. To determine L^*, note that $P[L^* \le \rho] = P[0 \le \delta]$ where $\delta = \sum_{q=1}^{P} c_q \theta_q - L^* \sum_{r=P+1}^{Q} c_r \theta_r$. For any given value of L^*, (3.3.3) can be used to construct an upper interval on δ with confidence coefficient close to $1 - \alpha$. Thus, to ensure $P[L^* \leq \rho]$ is close to $1 - \alpha$, we set $L = 0$ where L is defined in (3.3.3) and solve the resulting quadratic equation for L^*. The resulting lower bound for an upper $1 - \alpha$ interval on ρ is

$$L^* = \hat{\rho} \left[\frac{2 + k_4/(k_1 k_2) - \sqrt{V_L^*}}{2(1 - k_5/k_2^2)} \right]$$

(3.4.3)

where

$$V_L^* = (2 + k_4/(k_1 k_2))^2 - 4(1 - k_5/k_2^2)(1 - k_3/k_1^2)$$

$$k_1 = \sum_{q=1}^{P} c_q S_q^2 \qquad k_2 = \sum_{r=P+1}^{Q} c_r S_r^2$$

$$k_3 = \sum_{q=1}^{P} G_q^2 c_q^2 S_q^4 + \sum_{q=1}^{P-1} \sum_{t>q}^{P} G_{qt}^* c_q c_t S_q^2 S_t^2$$

$$k_4 = \sum_{q=1}^{P} \sum_{r=P+1}^{Q} G_{qr} c_q c_r S_q^2 S_r^2 \qquad k_5 = \sum_{r=P+1}^{Q} H_r^2 c_r^2 S_r^4$$

$\hat{\rho}$ is defined in (3.4.2) and G_q, H_r, G_{qr}, and G_{qt}^* are defined in (3.3.3). The upper $1 - \alpha$ bound derived in this same manner is

$$U^* = \hat{\rho} \left[\frac{2 + k_7/(k_1 k_2) + \sqrt{V_U^*}}{2(1 - k_8/k_2^2)} \right]$$

(3.4.4)

where

$$V_U^* = (2 + k_7/(k_1 k_2))^2 - 4(1 - k_8/k_2^2)(1 - k_6/k_1^2)$$

$$k_6 = \sum_{q=1}^{P} H_q^2 c_q^2 S_q^4 \qquad k_7 = \sum_{q=1}^{P} \sum_{r=P+1}^{Q} H_{qr} c_q c_r S_q^2 S_r^2$$

$$k_8 = \sum_{r=P+1}^{Q} G_r^2 c_r^2 S_r^4 + \sum_{r=P+1}^{Q-1} \sum_{u>r}^{Q} H_{ru}^* c_r c_u S_r^2 S_u^2 \qquad \text{and}$$

H_q, G_r, H_{qr}, and H_{ru}^* are defined in (3.3.4).

The bounds in (3.4.3) and (3.4.4) have several desirable properties. In particular, the confidence coefficients are exact when either

Table 3.4.1 Simulated Ranges of 95% Confidence Coefficients for One-Sided Intervals on $\rho = \Sigma_{q=1}^{P} c_q \theta_q / \Sigma_{r=P+1}^{Q} c_r \theta_r$ Using Satterthwaite (S) from (3.4.2) and Ting et al. (T) from (3.4.3) and (3.4.4)

P	Q	Numerator degrees of freedom	Denominator degrees of freedom	Lower intervals [0;U] S	T
2	3	5,5	5	.938–.954	.946–.970
2	3	5,30	30	.908–.952	.949–.969
2	4	2,4	2,4	.915–.967	.944–.986
2	5	2,49	4,4,4	.849–.968	.940–.975
3	4	2,3,6	12	.889–.969	.950–.993
3	4	5,10,15	15	.923–.957	.943–.974
3	6	2,5,10	2,5,10	.870–.968	.944–.989
4	5	5,10,10,25	50	.902–.959	.950–.983
4	7	6,6,6,36	3,12,12	.915–.972	.944–.985

P	Q	Numerator degrees of freedom	Denominator degrees of freedom	Upper intervals [L;∞) S	T
2	3	5,5	5	.950–.955	.950–.954
2	3	5,30	30	.948–.960	.948–.954
2	4	2,4	2,4	.915–.968	.948–.985
2	5	2,49	4,4,4	.922–.973	.944–.994
3	4	2,3,6	12	.950–.962	.948–.956
3	4	5,10,15	15	.949–.957	.947–.953
3	6	2,5,10	2,5,10	.870–.971	.941–.989
4	5	5,10,10,25	50	.947–.963	.941–.953
4	7	6,6,6,36	3,12,12	.874–.963	.940–.986

$P = 1$ and $Q = 2$, or when all but one of the degrees of freedom approach infinity. If $k_3/k_1^2 + k_5/k_2^2 + k_4/(k_1 k_2) \geq 0$ and $k_6/k_1^2 + k_8/k_2^2 + k_7/(k_1 k_2) \geq 0$, then (i) $V_U^* \geq 0$, (ii) $V_L^* \geq 0$, and (iii) $0 \leq L^* \leq \hat{\rho} \leq U^*$. As discussed in Ting, Burdick, and Graybill (1991), these inequality conditions are generally met for the values of α, P, and Q used in practice. If a situation should arise where either (i) or (ii) is violated, (3.4.2) can be used to construct an interval on ρ. The bounds in (3.4.3) and (3.4.4) also have a symmetric relationship. If U^* is a $1 - \alpha$ upper bound on ρ, and L^* is a $1 - \alpha$ lower bound on $1/\rho$, then $1/U^* = L^*$.

Ting, Burdick, and Graybill (1991) used computer simulation to compare (3.4.3) and (3.4.4) with the bounds in (3.4.2) for designs where Q ranged from 3 to 8 and P from 2 to 4. A summary of these results is shown in Tables 3.4.1 and 3.4.2. Table 3.4.1 provides simulated confidence coefficients based on 10,000 iterations over feasible ranges of the $\theta_i/\Sigma_{i=1}^{Q} \theta_i$. If the true confidence coefficient is 90% (95%), there is less than a 2.5% chance that a simulated confidence coefficient will be less than 89.4% (94.6%).

Ting et al. concluded that (3.4.3) and (3.4.4) are preferred over the Satterthwaite bounds because they better maintain the stated confidence level. As shown in Table 3.4.1, the simulated confidence coefficients of one-sided 95% Satterthwaite intervals drop as low as 87.0% for upper intervals (lower bounds) and as low as 84.9% for lower intervals (upper bounds). The corresonding values for (3.4.3) and (3.4.4) are both 94.0%. Table 3.4.2 shows that two-sided Satterthwaite intervals with stated confidence coefficients of 90% drop as low as 81.1% whereas those based on (3.4.3) and (3.4.4) never drop below 89.2%. Birch, Burdick, and Ting (1990) also compared (3.4.3) and

Table 3.4.2 Simulated Ranges of 90% Confidence Coefficients for Two-Sided Intervals on $\rho = \Sigma_{q=1}^{P} c_q \theta_q / \Sigma_{r=P+1}^{Q} c_r \theta_r$ Using Satterthwaite (S) from (3.4.2) and Ting et al. (T) from (3.4.3) and (3.4.4)

P	Q	Numerator degrees of freedom	Denominator degrees of freedom	Two-sided intervals $[L;U]$ S	T
2	3	5,5	5	.892–.908	.898–.922
2	3	5,30	30	.864–.906	.897–.923
2	4	2,4	2,4	.857–.934	.900–.970
2	5	2,49	4,4,4	.811–.934	.895–.963
3	4	2,3,6	12	.849–.926	.903–.944
3	4	5,10,15	15	.878–.908	.893–.923
3	6	2,5,10	2,5,10	.816–.930	.893–.974
4	5	5,10,10,25	50	.863–.914	.898–.930
4	7	6,6,6,36	3,12,12	.835–.924	.892–.959

(3.4.4) to a modification of (3.4.2) for the special case $P = 2, Q = 4$ and found (3.4.3) and (3.4.4) to be superior.

Other intervals for special cases of ρ have been proposed. Lu, Graybill, and Burdick (1987) and Birch and Burdick (1989) developed intervals for the cases where $P = 2, Q = 3$ and $P = 3, Q = 4$, respectively. Jeyaratnam and Graybill (1980) derived an interval on ρ for the random three-factor crossed design in which $P = 2$ and $Q = 4$. There is not much practical difference between these intervals and those formed with (3.4.3) and (3.4.4).

When intervals are desired for ρ^{-1}, appropriate transformations can be performed on intervals for ρ. For example, if $[L;U]$ is a two-sided $1 - \alpha$ interval on ρ, then $[1/U;1/L]$ is a $1 - \alpha$ interval on ρ^{-1}.

3.4.3   Intervals on Other Ratios

The coefficients for the general ratio considered in Section 3.4.2 were all nonnegative. However, negative coefficients are sometimes required to define a ratio of interest. Although no single method has been proposed for constructing intervals on all ratios of this form, results for one special case have been obtained.

Wang and Graybill (1981) derived upper and lower confidence intervals on the ratio $\eta = (c_1\theta_1 - c_2\theta_2)/\theta_3$ where $c_1,c_2 \geq 0$ and it is known $\eta \geq 0$. This parameter is used, for example, to represent σ_A^2/σ_E^2 in the two-fold nested design. In this model $\theta_1 = \sigma_E^2 + K\sigma_B^2 + JK\sigma_A^2$, $\theta_2 = \sigma_E^2 + K\sigma_B^2, \theta_3 = \sigma_E^2$, and $\sigma_A^2/\sigma_E^2 = (\theta_1 - \theta_2)/(JK\theta_3)$. A lower confidence interval on η was also proposed by Wang (1978).

Lu, Graybill, and Burdick (1989) extended the results of Wang and Graybill (1981) and Wang (1978) to the case where the sign of η is unknown. Such a situation arises, for example, in a two-factor crossed design if one is interested in the ratio $(\sigma_A^2 - \sigma_B^2)/\sigma_E^2$. Lu, Graybill, and Burdick also proposed a third method based on an interval for $\delta = c_1\theta_1 - c_2\theta_2$ proposed by Lu, Graybill, and Burdick (1988). Based on the Lu, Graybill, and Burdick (1989) paper, we recommend the following guidelines for constructing intervals on η.

If $\eta \geq 0$, the method developed by Wang and Graybill (1981) is recommended for both one-sided and two-sided intervals. The lower bound of an upper $1 - \alpha$ interval is

$$L = \frac{c_2 S_2^2}{S_3^2 \, F_{\alpha:n_1,n_3}} \left[T - F_{\alpha:n_1,\infty} + \frac{F_{\alpha:n_1,n_2} \, (F_{\alpha:n_1,\infty} - F_{\alpha:n_1,n_2})}{T} \right] \quad (3.4.5)$$

where

$$T = \frac{c_1 S_1^2}{c_2 S_2^2}$$

If $T \le F_{\alpha:n_1,n_2}$, L is defined to be zero.
The upper bound of a lower $1 - \alpha$ interval is

$$U = \frac{c_2 S_2^2}{S_3^2 \, F_{1-\alpha:n_1,n_3}} \left[T - F_{1-\alpha:n_1,\infty} + \frac{F_{1-\alpha:n_1,n_2} \, (F_{1-\alpha:n_1,\infty} - F_{1-\alpha:n_1,n_2})}{T} \right] \quad (3.4.6)$$

If $T \le F_{1-\alpha:n_1,n_2}$ then U is defined to be zero.

If the sign of η is unknown, the theorem developed by Lu, Graybill, and Burdick (1989) can be used to extend the results of (3.4.5) and (3.4.6). However, the upper intervals that result can sometimes be very liberal. Two alternative sets of bounds were examined by Lu, Graybill, and Burdick (1989) and found to perform better than (3.4.5) and (3.4.6) when the sign of η is unknown. Both sets of bounds provide intervals that maintain stated confidence levels. We present the bounds that appear to perform better in nested and crossed designs. These bounds were derived by extending results proposed by Howe (1974). The recommended lower bound for an upper $1 - \alpha$ interval on η when the sign is unknown is

$$L = \begin{cases} c_0 \hat{\eta} - \dfrac{\sqrt{V_{L1}}}{S_3^2} & \text{if } T \le F_{\alpha:n_1,n_2} \\[2ex] c_0 \hat{\eta} - \dfrac{\sqrt{V_{L2}}}{S_3^2} & \text{if } T > F_{\alpha:n_1,n_2} \end{cases}$$

where

$$\hat{\eta} = \frac{c_1 S_1^2 - c_2 S_2^2}{S_3^2} \qquad c_0 = 1 - \frac{2}{n_3} \qquad (3.4.7)$$

$$V_{L1} = \left[c_0^2 \left(1 - \frac{1}{F_{\alpha:n_1,n_2}} \right)^2 - \left(\frac{\frac{1}{F_{1-\alpha:n_2,n_3}} - c_0}{F_{\alpha:n_1,n_2}} \right)^2 \right] c_1^2 S_1^4$$

$$+ \left(\frac{1}{F_{1-\alpha:n_2,n_3}} - c_0 \right)^2 c_2^2 S_2^4 \qquad \text{and}$$

$$V_{L2} = \left[\left(c_0 - \frac{1}{F_{\alpha:n_1,n_3}} \right)^2 c_1^2 S_1^4 \right.$$

$$+ \left. \left[c_0^2 (F_{\alpha:n_1,n_2} - 1)^2 - F_{\alpha:n_1,n_2}^2 \left(c_0 - \frac{1}{F_{\alpha:n_1,n_3}} \right)^2 \right] c_2^2 S_2^4 \right]$$

The upper bound for a lower $1 - \alpha$ interval on η with unknown sign is

$$U = \begin{cases} c_0 \hat{\eta} + \dfrac{\sqrt{V_{U1}}}{S_3^2} & \text{if } T \leq F_{1-\alpha:n_1,n_2} \\ c_0 \hat{\eta} + \dfrac{\sqrt{V_{U2}}}{S_3^2} & \text{if } T > F_{1-\alpha:n_1,n_2} \end{cases}$$

where

$$V_{U1} = \left[c_0^2 \left(\frac{1}{F_{1-\alpha:n_1,n_2}} - 1 \right)^2 - \left(\frac{c_0 - \frac{1}{F_{\alpha:n_2,n_3}}}{F_{1-\alpha:n_1,n_2}} \right)^2 \right] c_1^2 S_1^4$$

$$+ \left(c_0 - \frac{1}{F_{\alpha:n_2,n_3}} \right)^2 c_2^2 S_2^4 \qquad \text{and} \qquad (3.4.8)$$

$$V_{U2} = \left[\left(\frac{1}{F_{1-\alpha:n_1,n_3}} - c_0 \right)^2 c_1^2 S_1^4 \right.$$

$$\left. + \left[c_0^2 (1 - F_{1-\alpha:n_1,n_2})^2 - F_{1-\alpha:n_1,n_2}^2 \left(\frac{1}{F_{1-\alpha:n_1,n_3}} - c_0 \right)^2 \right] c_2^2 S_2^4 \right]$$

The bounds (3.4.7) and (3.4.8) are exact when either (i) $c_1\theta_1 = 0$, or (ii) $c_2\theta_2 = 0$, or (iii) $c_1\theta_1 = c_2\theta_2$.

We have presented in this section general ratio forms for which good approximate intervals exist. In later chapters we will present other special forms that are unique to particular experimental designs. For most of these ratios, certain expected mean squares appear in both the numerator and denominator and there are order constraints on the θ_q. For example, in Chapter 6 we present a method developed by Ting and Graybill (1991) for setting a confidence interval on $(\theta_1 - \theta_3)/(\theta_2 - \theta_3)$ where $\theta_1 \geq \theta_3$ and $\theta_2 \geq \theta_3$. The final section of this chapter presents results for constructing simultaneous intervals on functions of variance components in balanced designs.

3.5   SIMULTANEOUS CONFIDENCE INTERVALS

As mentioned in Section 2.4, investigators often desire an overall protection level for a set of confidence intervals based on the same sample. Hartley and Rao (1967) derived an inequality based on the F-distribution that forms a confidence region on the set of ratios θ_q/θ_Q for $q = 1, \ldots, Q - 1$. This region also can be computed for mixed and unbalanced models. However, the confidence region is defined in a $Q - 1$ dimensional space and is difficult to interpret. If hypothesis testing is the major interest, one can determine if a hypothesized vector is in the region, although computations are complex.

Broemeling (1969a,b) considered balanced random models and used Kimball's inequality with (3.4.1) to set simultaneous one-sided confidence intervals on θ_q/θ_Q for $q = 1, \ldots, Q - 1$. The qth upper bound in the set of lower intervals is

$$U_q = \frac{S_q^2}{S_Q^2 \, F_{1-\alpha_q:n_q,n_Q}} \qquad q = 1, \ldots, Q - 1 \qquad (3.5.1)$$

The confidence coefficient for the simultaneous set of lower intervals, $1 - \alpha$, is greater than or equal to $\Pi_{q=1}^{Q-1} (1 - \alpha_q)$. A set of upper intervals can be established using (3.5.1) and replacing $F_{1-\alpha_q:n_q,n_Q}$ with $F_{\alpha_q:n_q,n_Q}$ values. Broemeling (1978) extended these one-sided interval results to two-sided intervals, but Tong (1979) showed the two-sided results do not always hold true. The two-sided inequalities proposed by Broemeling (1978) are true, however, when $n_1 = \cdots = n_{Q-1}$ and $\alpha_1 = \cdots = \alpha_{Q-1}$. Additionally, they hold true in the general case of unequal n_q when the probabilities $(1 - \alpha_q)$ are all large.

Khuri (1981) developed a method for constructing simultaneous two-sided confidence intervals on all continuous functions of variance components in balanced random models. Using (3.2.1), a two-sided $1 - 2\alpha_K$ interval on θ_q $(q = 1, \ldots, Q)$ is

$$\left[\frac{S_q^2}{F_{\alpha_K:n_q,\infty}} \, ; \, \frac{S_q^2}{F_{1-\alpha_K:n_q,\infty}} \right] \qquad (3.5.2)$$

Since the S_q^2 are independent, a rectangular confidence region on the θ_q with confidence coefficient $1 - 2\alpha = (1 - 2\alpha_K)^Q$ is defined by the Cartesian product of the Q confidence intervals in (3.5.2). We represent this confidence region as the set A.

In many applications, constraints are placed on the θ_q. For example, in Table 2.2.2 note that $\theta_2 \geq \theta_3 \geq \theta_4$. It is desirable, therefore, for a confidence region to include only points in the space of vectors that satisfy any existing constraints. We denote this constrained space as R. In situations where A is not completely contained in R, optimization programs are needed to compute the confidence regions proposed by Khuri. However, if A is contained in R, the confidence region can be written in closed form. We present results for this special case.

Khuri (1981) provides results for constructing simultaneous intervals on all linear combinations of the form $\delta = \Sigma_{q=1}^{P} c_q \theta_q - \Sigma_{r=P+1}^{Q} c_r \theta_r$ where $c_q, c_r \geq 0$. These intervals are conservative with an overall confidence coefficient of at least $1 - 2\alpha$. If $A \subset R$, these simultaneous lower and upper bounds are

$$L = \sum_{q=1}^{P} \frac{c_q S_q^2}{F_{\alpha_K:n_q,\infty}} - \sum_{r=P+1}^{Q} \frac{c_r S_r^2}{F_{1-\alpha_K:n_r,\infty}} \quad \text{and} \quad (3.5.3)$$

$$U = \sum_{q=1}^{P} \frac{c_q S_q^2}{F_{1-\alpha_K:n_q,\infty}} - \sum_{r=P+1}^{Q} \frac{c_r S_r^2}{F_{\alpha_K:n_r,\infty}}$$

where

$$\alpha_K = \frac{1 - (1 - 2\alpha)^{1/Q}}{2}$$

Khuri also provides a simultaneous confidence region with confidence coefficient at least $1 - 2\alpha$ on all ratios of the form δ/θ_Q. If $A \subset R$, the lower and upper bounds can be written as

$$L_R = \frac{LF_{1-\alpha_K:n_Q,\infty}}{S_Q^2} \quad \text{and} \quad U_R = \frac{UF_{\alpha_K:n_Q,\infty}}{S_Q^2} \quad (3.5.4)$$

where α_K, L, and U are defined in (3.5.3).

It is of interest to compare (3.5.4) with the bounds proposed by Broemeling. Consider the case where one desires simultaneous lower intervals (upper bounds) on θ_q/θ_Q for ($q = 1, ..., Q - 1$) with no constraints on the θ_q so that $A \subset R$. Assume one desires an overall confidence of $1 - \alpha$ with equal confidence coefficients for each individual interval. The upper bounds for these intervals can be computed using either (3.5.1) or U_R in (3.5.4). For this problem, (3.5.1) provides tighter intervals if

$$\frac{1}{F_{1-\alpha_B:n_q,n_Q}} < \frac{F_{\alpha_K:n_Q,\infty}}{F_{1-\alpha_K:n_q,\infty}} \quad (3.5.5)$$

where

$$\alpha_K = 1 - (1 - \alpha)^{\frac{1}{Q}} \quad \text{and} \quad \alpha_B = 1 - (1 - \alpha)^{\frac{1}{(Q-1)}}$$

For the case where $\alpha_B = \alpha_K = \alpha^*$, Burk et al. (1984) have shown (3.5.5) is always true if $n_q \leq n_Q$. If $n_q > n_Q$, it is not true for all α^*, but is always true for the small values of α^* associated with confidence intervals. Given this result and the fact that $\alpha_B > \alpha_K$, (3.5.5) must be true. A similar argument can be made for lower bounds in upper intervals. Thus, if one-sided simultaneous intervals on θ_q/θ_Q are desired, the intervals proposed by Broemeling are recommended over those proposed by Khuri. The same recommendation is made for two-sided simultaneous intervals on θ_q/θ_Q when $1 - \alpha_B$ is large.

Simultaneous intervals on functions of variance components can also be obtained using the Bonferroni inequality in (2.4.4) with the one-at-a-time intervals presented in this chapter. If the number of desired intervals is less than Q, this procedure produces shorter intervals than Khuri's method. The following example provides an illustration.

Example 3.5.1 Khuri (1981) reports a data set originally presented by Crump (1946) for a series of genetic experiments with fruit flys. The experimental design is the random two-factor crossed model shown in Table 2.2.2. The ANOVA table for the data is shown in Table 3.5.1.

Table 3.5.1 Analysis of Variance for Fruit Fly Data

SV	DF	MS	EMS
A (Experiments)	3	46,659	$\theta_1 = \sigma_E^2 + 12\sigma_{AB}^2 + 300\sigma_A^2$
B (Races)	24	3,243	$\theta_2 = \sigma_E^2 + 12\sigma_{AB}^2 + 48\sigma_B^2$
Interaction	72	459	$\theta_3 = \sigma_E^2 + 12\sigma_{AB}^2$
Error	1,100	231	$\theta_4 = \sigma_E^2$

To illustrate Khuri's method, we first compute a $1 - 2\alpha = .90$ simultaneous set of two-sided confidence intervals on $\theta_1, \theta_2, \theta_3$, and θ_4. Since $Q = 4$, an overall two-sided confidence of at least 90% is obtained using (3.5.2) with $\alpha_K = (1 - (.90)^{1/4})/2 = .013$. Using the function CINV of SAS©, we obtain $F_{.013:3,\infty} = 3.592$, $F_{.013:24,\infty} = 1.749$, $F_{.013:72,\infty} = 1.407$, $F_{.013:1100,\infty} = 1.097$, $F_{.987:3,\infty} = .0458$, $F_{.987:24,\infty} = .4693$, $F_{.987:72,\infty} = .6661$, and $F_{.987:1100,\infty} = .9075$. The computed confidence intervals on $\theta_1, \theta_2, \theta_3$, and θ_4 are [12,990; 1,018,756], [1854; 6910], [326; 689], and [211; 255], respectively. These four intervals form the confidence region A. For this model, the expected mean squares satisfy the constraints $\theta_1 \geq \theta_3 \geq \theta_4$ and $\theta_2 \geq \theta_3 \geq \theta_4$. This defines the constrained space, R. By looking at the computed intervals that form A, it is seen that A is completely contained in R. Thus, we can use (3.5.3) to compute confidence intervals on the variance components σ_A^2, σ_B^2, σ_{AB}^2, and σ_E^2. Since $\sigma_A^2 = (\theta_1 - \theta_3)/300$, the lower bound on σ_A^2 is computed using the lower bound on θ_1 and the upper bound on θ_3. That is, $L = (12,990 - 689)/300 = 41$. The upper bound on σ_A^2 is $U = (1,018,756 - 326)/300 = 3395$. In a similar manner, the confidence intervals on σ_B^2, σ_{AB}^2 and σ_E^2 are computed as [24.3; 137], [5.92; 39.8], and [211; 255], respectively.

Now suppose the investigator wants a set of 90% simultaneous two-sided confidence intervals on only σ_A^2, σ_B^2, and σ_{AB}^2. The Bonferroni inequality can be used with (3.3.1) and (3.3.2) to provide shorter intervals than those computed with Khuri's method. To illustrate, we compute the confidence interval on σ_A^2 using this approach. If we desire equal confidence coefficients on the individual intervals for σ_A^2, σ_B^2, and σ_{AB}^2, equation (2.4.4) requires that $1 - 6\alpha = .90$ and $\alpha = .10/6 = .01667$ for each interval. That is, each individual two-sided interval has a confidence coefficient of $1 - 2(.01667) = .967$. Since $\sigma_A^2 = (\theta_1 - \theta_3)/300$, we replace c_2, S_2^2, and n_2 with c_3, S_3^2, and n_3, respectively, in (3.3.1) and (3.3.2). Using CINV and FINV of SAS© we obtain $G_1 = .7069$, $H_3 = .4736$, $G_{13} = .03398$, $H_1 = 17.41$, $G_3 = .2788$, and $H_{13} = -1.239$. Placing these values into (3.3.1) and (3.3.2) with $S_1^2 = 46,659$, $S_3^2 = 459$, $n_1 = 3$, $n_3 = 72$, and $c_1 = c_3 = 1/300$, we compute $L = 44$ and $U = 2861$. This interval is shorter than the interval on σ_A^2 obtained from (3.5.3). Similar results are obtained for σ_B^2 and σ_{AB}^2.

Table 3.6.1 Summary Formulas for Confidence Intervals and MVU Estimators

Parameter	MVU Estimator	Confidence Interval
θ_q	S_q^2	(3.2.1)-Exact
$\gamma = \sum_q^Q c_q \theta_q$	$\sum_q^Q c_q S_q^2$	(3.2.5)-Two-sided interval (3.2.4)-Upper interval (3.2.5)-Lower interval
$\delta = \sum_{q=1}^P c_q \theta_q - \sum_{r=P+1}^Q c_r \theta_r$	$\sum_{q=1}^P c_q S_q^2 - \sum_{r=P+1}^Q c_r S_r^2$	(3.3.3),(3.3.4)
θ_1/θ_2	$\dfrac{S_1^2}{S_2^2}\left(1 - \dfrac{2}{n_2}\right)$	(3.4.1)-Exact
$\rho = \sum_{q=1}^P c_q \theta_q \Big/ \sum_{r=P+1}^Q c_r \theta_r$	—	(3.4.3),(3.4.4)
$\eta = (c_1\theta_1 - c_2\theta_2)/\theta_3$	$\dfrac{c_1 S_1^2 - c_2 S_2^2}{S_3^2}\left(1 - \dfrac{2}{n_3}\right)$	(3.4.5),(3.4.6) $\eta \geq 0$ (3.4.7),(3.4.8) (Sign unknown)

3.6 SUMMARY

This chapter presented confidence intervals for several functions of variance components in balanced random models. These intervals and MVU estimators are summarized in Table 3.6.1.

As will be illustrated in later chapters, these formulas can be used in unbalanced designs with appropriate sums of squares. The results in this chapter can also be used in mixed models as discussed in Chapter 7.

The methods presented in this chapter are recommended based on their success at maintaining stated confidence coefficients across a wide range of conditions examined in our research. However, if an investigator is concerned about the confidence coefficient in any particular application, we encourage the use of simulation to determine if the confidence coefficient is close enough to the nominal level.

4

The One-Fold Nested Design

4.1 INTRODUCTION

The model considered in this chapter is the random one-fold nested design. Although notationally simple, this model can be generalized to more complex designs and has proven useful to practitioners in a variety of fields. An example data set from a one-fold nested design is shown in Table 4.1.1. These data were reported by Swallow and Searle (1978) and represent the weights in ounces of five groups of bottles of vegetable oil that were selected at random from a moving production line. One objective of the experiment was to compare the variability in bottle weights among groups to the variability within groups.

The model for the one-fold nested design is

$$Y_{ij} = \mu + A_i + E_{ij} \qquad i = 1, \ldots, I; \quad j = 1, \ldots, J_i \qquad (4.1.1)$$

where μ is an unknown constant, A_i and E_{ij} are mutually independent normal random variables with means of zero and variances σ_A^2 and σ_E^2, respectively. For the data shown in Table 4.1.1, $I = 5, J_1 = 4, J_2 = 2, J_3 = 5, J_4 = 3$, and $J_5 = 2$. The analysis of variance for model

Table 4.1.1   Weights of Five Groups of Bottles in Ounces

		Group		
1	2	3	4	5
15.70	15.69	15.75	15.68	15.65
15.68	15.71	15.82	15.66	15.60
15.64		15.75	15.59	
15.60		15.71		
		15.84		
$J_1 = 4$	$J_2 = 2$	$J_3 = 5$	$J_4 = 3$	$J_5 = 2$
$\bar{Y}_{1*} = 15.655$	$\bar{Y}_{2*} = 15.700$	$\bar{Y}_{3*} = 15.774$	$\bar{Y}_{4*} = 15.643$	$\bar{Y}_{5*} = 15.625$

Table 4.1.2   Analysis of Variance for One-Fold Nested Design

SV	DF	SS[a]	MS	EMS
Among groups	$n_1 = I - 1$	SS1	$S_1^2 = SS1/n_1$	$\theta_1 = \sigma_E^2 + J_o\sigma_A^2$
Within groups	$n_2 = N - I$	SS2	$S_2^2 = SS2/n_2$	$\theta_2 = \sigma_E^2$
Total	$N - 1$	SST		

[a]$SS1 = \Sigma_i J_i (\bar{Y}_{i*} - \bar{Y}_{**})^2$, $SS2 = \Sigma_i\Sigma_j (Y_{ij} - \bar{Y}_{i*})^2$, and $SST = \Sigma_i\Sigma_j (Y_{ij} - \bar{Y}_{**})^2$

(4.1.1) is shown in Table 4.1.2 where $N = \Sigma_i J_i$, $\bar{Y}_{i*} = \Sigma_j Y_{ij}/J_i$, $\bar{Y}_{**} = \Sigma_i\Sigma_j Y_{ij}/N$, and $J_o = (N - \Sigma_i J_i^2/N)/(I - 1)$.

Formulas useful for hand computations are $SST = \Sigma_i\Sigma_j Y_{ij}^2 - Y_{**}^2/N$, $SS1 = \Sigma_i(Y_{i*}^2/J_i) - Y_{**}^2/N$, and $SS2 = SST - SS1$ where $Y_{i*} = \Sigma_j Y_{ij}$ and $Y_{**} = \Sigma_i\Sigma_j Y_{ij}$.

Several parameters are of interest in model (4.1.1). In addition to the variance components, σ_A^2 and σ_E^2, inferences are often desired for the parameters $\gamma = \sigma_A^2 + \sigma_E^2$, $\lambda_A = \sigma_A^2/\sigma_E^2$, $\lambda_E = \sigma_E^2/\sigma_A^2$, $\rho_A = \sigma_A^2$ and $\rho_E = \sigma_E^2/\gamma$. Section 4.2 presents methods for constructing confidence intervals on these parameters in the balanced case where all $J_i = J$.

Section 4.3 considers the same problem in unbalanced models where all J_i are not equal.

4.2   THE BALANCED ONE-FOLD RANDOM MODEL

Model (4.1.1) is balanced when $J_i = J$ for all I groups. In this situation, $J_o = J$ and $\theta_1 = \sigma_E^2 + J\sigma_A^2$. Additionally, $n_1 S_1^2/\theta_1$ and $n_2 S_2^2/\theta_2$ are independently distributed chi-squared random variables with n_1 and n_2 degrees of freedom, respectively. Thus, the results of Chapter 3 can be used to form confidence intervals on parameters of interest in this balanced design. These results are presented in this section.

4.2.1   Inferences on σ_E^2 and σ_A^2

For both the balanced and unbalanced one-fold designs, $n_2 S_2^2/\sigma_E^2$ has a chi-squared distribution with n_2 degrees of freedom. Confidence intervals on σ_E^2 are constructed using (3.2.1). The exact two-sided $1 - 2\alpha$ confidence interval on σ_E^2 is

$$\left[\frac{S_2^2}{F_{\alpha:n_2,\infty}} ; \frac{S_2^2}{F_{1-\alpha:n_2,\infty}} \right] \tag{4.2.1}$$

Interval (4.2.1) can be used to test $H_o : \sigma_E^2 = c$ against $H_a : \sigma_E^2 \neq c$ for any value of c. If c is not contained in the two-sided interval, the null hypothesis is rejected at the 2α level. The lower bound in (4.2.1) can be used to test $H_o : \sigma_E^2 \leq c$ against $H_a : \sigma_E^2 > c$ by rejecting H_o if the lower bound is greater than c. This provides a uniformly most powerful test. The upper bound in (4.2.1) can be used to test $H_o : \sigma_E^2 \geq c$ against $H_a : \sigma_E^2 < c$ by rejecting H_o if the upper bound is less than c. This provides a uniformly most powerful unbiased test (see, e.g., Lehmann (1986, p. 108–111, p. 192–197)). The MVU estimator for σ_E^2 is S_2^2.

By writing $\sigma_A^2 = (\theta_1 - \theta_2)/J$, it is seen that σ_A^2 is a special case of the parameter $\delta = c_1\theta_1 - c_2\theta_2$ discussed in Section 3.3.1. Accordingly, we recommend (3.3.1) and (3.3.2) for constructing confidence intervals on σ_A^2. Using the notation of this chapter, the resulting approximate two-sided $1 - 2\alpha$ confidence interval on σ_A^2 is

$$\left[\frac{S_1^2 - S_2^2 - \sqrt{V_L}}{J} \; ; \; \frac{S_1^2 - S_2^2 + \sqrt{V_U}}{J} \right] \qquad (4.2.2)$$

where

$$V_L = G_1^2 S_1^4 + H_2^2 S_2^4 + G_{12} S_1^2 S_2^2 \qquad V_U = H_1^2 S_1^4 + G_2^2 S_2^4 + H_{12} S_1^2 S_2^2$$

$$G_\ell = 1 - \frac{1}{F_{\alpha:n_\ell,\infty}} \quad (\ell = 1,2) \qquad H_\ell = \frac{1}{F_{1-\alpha:n_\ell,\infty}} - 1 \quad (\ell = 1,2)$$

$$G_{12} = \frac{(F_{\alpha:n_1,n_2} - 1)^2 - G_1^2 F_{\alpha:n_1,n_2}^2 - H_2^2}{F_{\alpha:n_1,n_2}} \qquad \text{and}$$

$$H_{12} = \frac{(1 - F_{1-\alpha:n_1,n_2})^2 - H_1^2 F_{1-\alpha:n_1,n_2}^2 - G_2^2}{F_{1-\alpha:n_1,n_2}}$$

The lower bound in (4.2.2) is negative if $S_1^2/S_2^2 < F_{\alpha:n_1,n_2}$ and the upper bound is negative if $S_1^2/S_2^2 < F_{1-\alpha:n_1,n_2}$. However, since $\sigma_A^2 \geq 0$, negative bounds are defined to be zero.

The decision rule for an exact size α test of $H_o : \sigma_A^2 = 0$ against $H_a : \sigma_A^2 > 0$ is to reject H_o if $S_1^2/S_2^2 > F_{\alpha:n_1,n_2}$. This is an exact test that is uniformly most powerful unbiased (see, e.g., Lehmann (1986, p. 418–422)). Alternatively, the test can be conducted using an upper $1 - \alpha$ confidence interval on σ_A^2. If the upper interval does not contain zero, H_o is rejected. Since the upper interval contains zero if and only if $S_1^2/S_2^2 < F_{\alpha:n_1,n_2}$, this approach provides the same result as the uniformly most powerful unbiased test. Approximate tests for $H_o : \sigma_A^2 = c$ against $H_a : \sigma_A^2 \neq c$, $H_o : \sigma_A^2 \geq c$ against $H_a : \sigma_A^2 < c$, and $H_o : \sigma_A^2 \leq c$ against $H_a : \sigma_A^2 > c$ where c is a specified positive constant, can be based on (4.2.2). The decision rule is to reject H_o if and only if the appropriate two-sided or one-sided confidence interval does not include c.

The MVU estimator for σ_A^2 is the ANOVA estimator $(S_1^2 - S_2^2)/J$. If $S_1^2/S_2^2 < 1$, the estimate is negative. If negative estimates are defined to be zero, the ANOVA estimator loses the property of unbiasedness, and hence MVU.

Example 4.2.1   Table 4.2.1 presents a data set from a balanced one-fold design with $I = 4$ and $J = 3$. These data represent fill weights of bottles selected at random from a sample of 4 filling machines.

The ANOVA table for these data is presented in Table 4.2.2. The computational formulas for the sums of squares provide $SST = (14.23^2 + \cdots + 14.91^2) - (184.83)^2/12 = 6.784$, $SS1 = (44.04^2 + \cdots + 46.92^2)/3 - (184.83)^2/12 = 5.329$, and $SS2 = 6.784 - 5.329 = 1.455$. From the sums of squares we compute $S_1^2 = 5.329/3 = 1.776$ and $S_2^2 = 1.455/8 = .1819$. Using equation

Table 4.2.1   Weights of Bottles Selected from Filling Machines

	\multicolumn{4}{c}{Machine}			
	1	2	3	4
	14.23	16.46	14.98	15.94
	14.96	16.74	14.88	16.07
	14.85	15.94	14.87	14.91
Y_{i*}	44.04	49.14	44.73	46.92
$\tilde{Y}_{i*}$	14.68	16.38	14.91	15.64

Table 4.2.2   Analysis of Variance for Data in Table 4.2.1

SV	DF	SS	MS
Among machines	3	5.329	1.776
Within machines	8	1.455	.1819
Total	11	6.784	

(4.2.1) the two-sided 95% confidence interval on σ_E^2 is $[.1819/2.1918;$ $.1819/.2725] = [.083; .668]$. The two-sided 95% confidence interval on σ_A^2 is computed using (4.2.2) with $J = 3$, $G_1 = .679$, $H_2 = 2.67$, $G_{12} = -.213$, $H_1 = 12.9$, $G_2 = .544$, and $H_{12} = -3.14$. The computed interval is $[.107; 8.16]$. The MVU estimate for σ_A^2 is $(1.776-.1819)/3 = .531$.

4.2.2 Inferences on $\gamma = \sigma_A^2 + \sigma_E^2$

The total variation of the response variable in the one-fold nested design is $\gamma = \sigma_A^2 + \sigma_E^2$. In terms of our general notation, $\gamma = \Sigma_q^Q c_q \theta_q$ where $Q = 2$, $\theta_1 = \sigma_E^2 + J\sigma_A^2$, $\theta_2 = \sigma_E^2$, $c_1 = 1/J > 0$, and $c_2 = 1 - 1/J > 0$. Thus, the results of Section 3.2 can be applied to this problem.

In particular, the recommended method for constructing a two-sided confidence interval on γ is the method proposed by Graybill and Wang (1980). Unlike the Satterthwaite and Welch approximations, this method provides an interval with a confidence coefficient that is close to $1 - 2\alpha$ and in most cases is conservative. The Graybill-Wang two-sided $1 - 2\alpha$ confidence interval on γ is reported in (3.2.5). In the notation of the one-fold nested design this interval is

$$\left[\hat{\gamma} - \frac{\sqrt{G_1^2 S_1^4 + G_2^2 (J - 1)^2 S_2^4}}{J} ; \hat{\gamma} + \frac{\sqrt{H_1^2 S_1^4 + H_2^2 (J - 1)^2 S_2^4}}{J} \right] \quad (4.2.3)$$

where

$$\hat{\gamma} = \frac{S_1^2 + (J - 1)S_2^2}{J} \qquad G_\ell = 1 - \frac{1}{F_{\alpha:n_\ell,\infty}} \quad (\ell = 1,2)$$

$$\text{and} \qquad H_\ell = \frac{1}{F_{1-\alpha:n_\ell,\infty}} - 1 \quad (\ell = 1,2)$$

Note the definitions of G_1, G_2, H_1, and H_2, are identical to those in (4.2.2). As discussed in Section 3.2.4, not only does (4.2.3) maintain its stated confidence level, but in many cases it will be shorter than the corresponding Welch interval.

If one-sided intervals on γ are desired, we recommend the upper $1 - \alpha$ Welch interval (lower bound) shown in (3.2.4). Although this bound is sometimes liberal, it is always very close to the stated level.

The recommended lower $1 - \alpha$ interval (upper bound) is the Graybill-Wang upper bound shown in (4.2.3).

Approximate tests for $H_o : \gamma = c$ against $H_a : \gamma \neq c$ can be conducted using interval (4.2.3). The recommended one-sided intervals can be used for testing one-sided alternatives. The MVU estimator for γ, $\hat{\gamma}$, is defined in (4.2.3).

Example 4.2.2   For the data in Example 4.2.1, the MVU estimate of γ is $1.776/3 + (2/3)(.1819) = .713$. Using (4.2.3) with $G_1 = .679, G_2 = .544, H_1 = 12.9$, and $H_2 = 2.67$, we compute the 95% interval on γ to be $[.306; 8.36]$.

4.2.3   Inferences on Ratios of Variance Components

The ratio $\rho_A = \sigma_A^2/(\sigma_A^2 + \sigma_E^2)$ is referred to as the intraclass correlation. This ratio is of interest in many fields including epidemiology, genetics, quality control, and tests and measurement. This ratio represents the percentage of total variation in a response variable accounted for by the factor under investigation. In the context of model (4.1.1), it can also be viewed as the correlation between any two observations in the same group.

The intraclass correlation can be expressed in alternative forms to define other ratios of the variance components. For example, $1 - \rho_A = \sigma_E^2/(\sigma_A^2 + \sigma_E^2) = \rho_E$, $\rho_A/(1 - \rho_A) = \sigma_A^2/\sigma_E^2 = \lambda_A$, and $(1 - \rho_A)/\rho_A = \sigma_E^2/\sigma_A^2 = \lambda_E$.

A confidence interval on ρ_A is derived from a confidence interval on λ_A. An exact interval on $1 + J\lambda_A$ is formed using S_1^2 and S_2^2 in equation (3.4.1). By appropriate transformation of this interval, an exact two-sided $1 - 2\alpha$ confidence interval on λ_A is

$$\left[\frac{L^* - 1}{J} ; \frac{U^* - 1}{J} \right] \qquad (4.2.4)$$

where

$$L^* = \frac{S_1^2}{S_2^2 F_{\alpha:n_1,n_2}} \quad \text{and} \quad U^* = \frac{S_1^2}{S_2^2 F_{1-\alpha:n_1,n_2}}$$

Since $\lambda_A \geq 0$, negative bounds are defined to be zero.

Using (4.2.4) and noting $\rho_A = \lambda_A/(1 + \lambda_A)$, an exact $1 - 2\alpha$ confidence interval on ρ_A is

$$\left[\frac{L^* - 1}{L^* - 1 + J} ; \frac{U^* - 1}{U^* - 1 + J} \right] \qquad (4.2.5)$$

where L^* and U^* are defined in (4.2.4). Since $\rho_A \geq 0$, negative bounds are defined to be zero. Interval (4.2.5) can be used to test any two-sided alternative hypothesis concerning ρ_A. One-sided intervals can be used for one-sided alternatives. Donner and Eliasziw (1987) have derived power contours that can be used for determining sample size for one-sided tests. Singhal (1987) and Groggel, Wackerly, and Rao (1988) provide approximate confidence intervals on ρ_A under the assumption of non-normality. Confidence intervals on λ_E and ρ_E are obtained by appropriate transformations of (4.2.4) and (4.2.5).

The MVU estimator for λ_A is $[(S_1^2/S_2^2)(1 - 2/(N - I)) - 1]/J$. The MVU estimator for ρ_A was derived by Olkin and Pratt (1958), but cannot be written in closed form. Donoghue and Collins (1990) give a subroutine that can be used to calculate the MVU estimator. The ANOVA estimator formed by replacing σ_A^2 and σ_E^2 with their MVU estimators is $(S_1^2 - S_2^2)/[S_1^2 + (J - 1) S_2^2]$. This estimator is biased and can produce negative estimates. Donner (1986) provides a review of other estimators for ρ_A in both balanced and unbalanced designs. Multivariate extensions are presented by Konishi and Khatri (1990).

Example 4.2.3 The exact 95% confidence interval on λ_A for the data in Example 4.2.1 is [.268; 47.0]. This interval was obtained using (4.2.4) with $F_{.025;3,8} = 5.416$, $F_{.975;3,8} = .0688$, $L^* = 1.80$, and $U^* = 142$. The 95% interval on ρ_A using equation (4.2.5) is [.211; .979]. The MVU estimate for λ_A is $[(1.776/.1819)(1-2/8)-1]/3 = 2.11$. The ANOVA estimate for ρ_A is $(1.776-.1819)/[1.776+2(.1819)] = .745$.

4.2.4 Comparing Total Variability of Two Independent One-Fold Nested Designs

A problem of interest in some fields of application is the comparison of total variability of two independent populations. As an example, consider two brands of machines used to fill bottles with vegetable oil. Several machines of each brand are used to fill bottles and the bottles are weighed. The purpose of the experiment is to determine whether the bottle weight variability is different for the two machine brands. The experimental model is represented using (4.1.1) as

$$Y_{mij} = \mu_m + A_{mi} + E_{mij} \qquad (4.2.6)$$

where Y_{mij} represents the weight of the jth bottle filled on the ith machine of brand m, μ_m is a constant representing the average bottle weight for brand m, A_{mi} and E_{mij} are mutually independent normal random variables with zero means and variances σ_{mA}^2 and σ_{mE}^2, respectively, $m = 1,2$; $i = 1, \ldots, I_m$; and $j = 1, \ldots, J_m$. In the context of the example, I_m represents the number of machines that are tested for brand m, and J_m represents the number of bottles filled by each machine for brand m. The variance component σ_{mA}^2 can be viewed as a measure of bottle weight variability across machines of brand m and σ_{mE}^2 as a measure of variability for any particular machine of brand m. The total process variance for brand m is $\gamma_m = [\theta_{m1} + (J_m - 1)\theta_{m2}]/J_m = \sigma_{mA}^2 + \sigma_{mE}^2$. The problem of interest is to test $H_o : \gamma_1 = \gamma_2$. This can be done by constructing confidence intervals on either $\gamma_1 - \gamma_2$ or γ_1/γ_2. In the former case the hypothesis of interest is rejected if zero is not contained in the interval. In the latter case, an interval not containing 1 leads to rejection of the null hypothesis. A test based on the confidence interval computed for $\gamma_1 - \gamma_2$ using (3.3.3) and (3.3.4) leads to the same result as a test based on the interval for γ_1/γ_2 using (3.4.3) and (3.4.4). This is because the lower bound in (3.3.3) is positive and leads to the rejection of H_o if and only if the lower bound in (3.4.3) is greater than 1. Similarly, the upper bound in (3.3.4) is negative and leads to the rejection of H_o if and only if (3.4.4) is less than one. Birch et al. (1990) compared the power and size of this test to other alternatives and found this to be the preferred method.

Example 4.2.4 Consider the example given in this section and assume that a company is using brand 1 machines and will consider changing to brand 2 only if evidence suggests that bottle weight variability will be decreased. In the context of a hypothesis test, evidence for a change in brand is provided if the null hypothesis $H_o : \gamma_1 \leq \gamma_2$ is rejected in favor of the alternative $H_a : \gamma_1 > \gamma_2$. In order to perform this test, we compute a lower bound on γ_1/γ_2 and reject the null hypothesis if the lower bound is greater than one. Table 4.2.3 presents data from an experiment with $I_1 = I_2 = 4$ machines for each brand and $J_1 = J_2 = 6$ bottles filled with each machine.

Using (3.4.3) with $\alpha = .05$ and letting $\theta_1 = \theta_{11}, \theta_2 = \theta_{12}, \theta_3 = \theta_{21}$, and $\theta_4 = \theta_{22}$, we have $\gamma_1/\gamma_2 = (\theta_1 + 5\theta_2)/(\theta_3 + 5\theta_4)$ with $n_1 = n_3 = 3$, and $n_2 = n_4 = 20$. The data are used to calculate $S_1^2 = .002694$, $S_2^2 = .000358$, $S_3^2 = .000438$, $S_4^2 = .000108$, $\hat{\rho} = (S_1^2 + 5S_2^2)/(S_3^2 + 5 S_4^2) = 4.58$, $k_3/k_1^2 = .187$, $k_5/k_2^2 = 11.6$, $k_4/(k_1k_2) = -.765$, and $L^* = 1.03$. Since the lower bound is greater than one, this suggests $\gamma_1 > \gamma_2$ and the company should consider using brand 2 machines. The same conclusion will be obtained if (3.3.3) is used to compute an interval on $\gamma_1 - \gamma_2$ in order to perform the test. This is because L^* greater than 1 implys that L in (3.3.3) is greater than 0.

4.2.5 An Alternative Model Formulation

The covariance structure of the observations in model (4.1.1) can be written as

$$
\begin{aligned}
\text{Cov}(Y_{ij}, Y_{i'j'}) &= \sigma_E^2 + \sigma_A^2 && i = i'; j = j' \\
&= \sigma_A^2 && i = i'; j \neq j' && (4.2.7) \\
&= 0 && i \neq i'
\end{aligned}
$$

Table 4.2.3 Weights of Bottles (in Ounces) for Two Machine Brands

			Brand					
			1				2	
Machine	1	2	3	4	5	6	7	8
	15.66	15.69	15.73	15.72	15.78	15.78	15.76	15.77
	15.66	15.71	15.68	15.73	15.80	15.76	15.80	15.80
	15.70	15.68	15.73	15.74	15.78	15.76	15.78	15.78
	15.70	15.72	15.71	15.74	15.79	15.77	15.78	15.78
	15.68	15.71	15.67	15.73	15.78	15.76	15.79	15.77
	15.70	15.72	15.72	15.75	15.79	15.77	15.78	15.78

Since σ_A^2 is positive, model (4.1.1) imposes a positive correlation among the observations within the same treatment group. Smith and Murray (1984) proposed the alternative covariance structure

$$
\begin{aligned}
\text{Cov}(Y_{ij},Y_{i'j'}) &= \sigma_E^2 + \delta_A & i &= i'; j = j' \\
&= \delta_A & i &= i'; j \neq j' \qquad (4.2.8) \\
&= 0 & i &\neq i'
\end{aligned}
$$

where $0 \leq |\delta_A| \leq \sigma_E^2 + \delta_A$. The covariance between observations in the same group, δ_A, is now unrestricted in sign. Smith and Murray provide a cattle breeding example where negative correlations are expected among weaning weights of sibling calves. Model (4.2.8) provides an alternative formulation to (4.1.1) that might be considered if the ANOVA estimator of σ_A^2 is negative.

The sums of squares for model (4.2.8) are the same ones shown in Table 4.1.2. In the balanced design, the expected mean squares for S_1^2 and S_2^2 are now $\sigma_E^2 + J\delta_A$ and σ_E^2, respectively. Thus, (4.2.2) can be used to form an interval on δ_A. As noted in Chapter 3, interval (4.2.2) is based on a method that does not require δ_A to be positive. One can test $H_o : \delta_A = 0$ against $H_a : \delta_A \neq 0$, $H_o : \delta_A \leq 0$ against $H_a : \delta_A > 0$, or $H_o : \delta_A \geq 0$ against $H_a: \delta_A < 0$ using (4.2.2) with the appropriate F-values.

4.3 THE UNBALANCED ONE-FOLD MODEL

In many applications, the J_i values in model (4.1.1) are not equal for all I groups. Table 4.3.1 presents one such unbalanced data set. These data appear in Snedecor and Cochran (1980, p. 247) and represent the percentages of conceptions to services for successive samples of six randomly sampled bulls. It is of interest to compare the variation among bulls in the population to the variation within bulls. The analysis of variance is shown in Table 4.3.2.

Design unbalancedness creates a problem concerning the distribution of $n_1 S_1^2/\theta_1$. In the balanced design where all $J_i = J$, $n_1 S_1^2/\theta_1$ and $n_2 S_2^2/\theta_2$ are independent chi-squared random variables with n_1 and n_2 degrees of freedom, respectively. In the unbalanced case, S_1^2 and S_2^2 are still independent and $n_2 S_2^2/\theta_2$ still has a chi-squared distribution with n_2

Table 4.3.1 Percentages of Conceptions to Services

		Bull			
1	2	3	4	5	6
46	70	52	47	42	35
31	59	44	21	64	68
37		57	70	50	59
62		40	46	69	38
30		67	14	77	57
		64		81	76
		70		87	57
					29
					60
$J_1 = 5$	$J_2 = 2$	$J_3 = 7$	$J_4 = 5$	$J_5 = 7$	$J_6 = 9$
$\bar{Y}_{1*} = 41.20$	$\bar{Y}_{2*} = 64.50$	$\bar{Y}_{3*} = 56.29$	$\bar{Y}_{4*} = 39.60$	$\bar{Y}_{5*} = 67.14$	$\bar{Y}_{6*} = 53.22$

Table 4.3.2 Analysis of Variance for Bull Data

SV	DF	SS	MS
Among bulls	5	3322.06	664.41
Within bulls	29	7200.34	248.29
Total	34	10522.40	

degrees of freedom. However, unless $\sigma_A^2 = 0$, $n_1 S_1^2/\theta_1$ no longer has a chi-squared distribution. Thus, the results in Section 4.2 are not strictly valid. We now present alternative methods that are recommended for this situation.

4.3.1 Inferences on σ_E^2 and σ_A^2

Since the distribution of $n_2 S_2^2/\theta_2$ is still chi-squared with n_2 degrees of freedom, the results concerning σ_E^2 in Section 4.2.1 are still valid. In particular, (4.2.1) is used to form an exact $1 - 2\alpha$ confidence interval on σ_E^2, and S_2^2 is the MVU estimator for σ_E^2.

The results presented in Section 4.2.1 for σ_A^2 are based on the assumption that $n_1 S_1^2/\theta_1$ has a chi-squared distribution with n_1 degrees of freedom. However, in an unbalanced design, $n_1 S_1^2/\theta_1$ has a chi-squared distribution if and only if $\sigma_A^2 = 0$. If it is known that σ_A^2 is close to zero, then treating $n_1 S_1^2/\theta_1$ as a chi-squared random variable may be appropriate. To form a confidence interval on σ_A^2 under this assumption, one could use (4.2.2) and replace J with $J_o = (N - \Sigma_i J_i^2/N)/(I - 1)$. However, if σ_A^2 is far from zero, this procedure might result in very liberal intervals.

As another option, Thomas and Hultquist (1978) derived an alternative statistic to $n_1 S_1^2/\theta_1$ that can be used for constructing confidence intervals in the unbalanced model. This statistic is $n_1 S_{1U}^2/\theta_{1U}$ where

$$S_{1U}^2 = \frac{\bar{J}_H \Sigma_i (\bar{Y}_{i*} - \bar{Y}_*^*)^2}{n_1}, \quad E(S_{1U}^2) = \theta_{1U} = \sigma_E^2 + \bar{J}_H \sigma_A^2$$

(4.3.1)

$$\bar{Y}_{i*} = \sum_j Y_{ij}/J_i, \quad \bar{Y}_*^* = \sum_i \bar{Y}_{i*}/I \quad \text{and} \quad \bar{J}_H = \frac{I}{\Sigma_i 1/J_i}$$

Computationally one can use $\Sigma_i(\bar{Y}_{i*} - \bar{Y}_*^*)^2 = \Sigma_i \bar{Y}_{i*}^2 - (\Sigma_i \bar{Y}_{i*})^2/I$. Example 4.3.1 demonstrates how $n_1 S_{1U}^2/\bar{J}_H$ can be computed using a software package.

The term $n_1 S_{1U}^2$ represents the unweighted sum of squares of the treatment means and $\bar{J}_H$ represents the harmonic mean of the J_i values. Rankin (1974) recommended these same statistics for testing hypotheses in an unbalanced fixed effects one-fold model. Thomas and Hultquist showed that the moment generating function of $n_1 S_{1U}^2/\theta_{1U}$ approaches that of a chi-squared random variable with n_1 degrees of freedom as all J_i approach a constant or if either $\lambda_A = \sigma_A^2/\sigma_E^2$ or all J_i approach infinity. They also provided simulations that indicate $n_1 S_{1U}^2/\theta_{1U}$ is well approximated by a chi-squared random variable except in cases where $\lambda_A < .25$ and the design is extremely unbalanced. In situations where the Thomas-Hultquist approximation works well, an interval on σ_A^2 can be constructed using (4.2.2) with S_{1U}^2 and $\bar{J}_H$ replacing S_1^2 and J, respectively.

In extremely unbalanced designs where $\lambda_A < .25$, the chi-squared approximation for $n_1 S_{1U}^2/\theta_{1U}$ is not good and the Thomas-Hultquist procedure can produce a liberal confidence interval. Although the approximation based on $n_1 S_1^2/\theta_1$ might work well in this range, it too can produce liberal intervals. A method that performs well over the entire range of λ_A was developed by Burdick and Eickman (1986). This approximate two-sided $1 - 2\alpha$ confidence interval on σ_A^2 is

$$\left[\frac{S_{1U}^2 L_m}{F_{\alpha:n_1,\infty}(1 + \bar{J}_H L_m)} \; ; \; \frac{S_{1U}^2 U_M}{F_{1-\alpha:n_1,\infty}(1 + \bar{J}_H U_M)} \right] \qquad (4.3.2)$$

where

$$L_m = \frac{S_{1U}^2}{\bar{J}_H S_2^2 F_{\alpha:n_1,n_2}} - \frac{1}{m} \qquad U_M = \frac{S_{1U}^2}{\bar{J}_H S_2^2 F_{1-\alpha:n_1,n_2}} - \frac{1}{M}$$

$$m = Min(J_1, J_2, \ldots, J_I) \qquad \text{and} \qquad M = Max(J_1, J_2, \ldots, J_I).$$

If $L_m < 0$, the lower bound in (4.3.2) is defined to be zero. Similarly, if $U_M < 0$, the upper bound is defined to be zero.

When all J_i are equal, (4.3.2) reduces to the interval on σ_A^2 proposed by Tukey (1951) and Williams (1962). Burdick and Eickman used computer simulation to show that (4.3.2) has a confidence coefficient that is generally at least as great as the stated value. Although (4.3.2) is more conservative than the Thomas-Hultquist interval, the average interval lengths of the two intervals never differed by more than 5% in the simulations reported by Burdick and Eickman.

The test of $H_o : \sigma_A^2 = 0$ against $H_a : \sigma_A^2 > 0$ is performed using the ratio S_1^2/S_2^2. This is the same test described in Section 4.2.1. The test is exact, but unlike in the balanced case, it is not uniformly most powerful unbiased. Singh (1987) provides expressions that can be used for computing the power of this F-test. Donner and Koval (1989) present results that indicate the F-test is more powerful than the likelihood ratio test unless I is large and the design is extremely unbalanced. Jeyaratnam and Othman (1985) provide an approximate test for this hypothesis when the variances of the E_{ij} are unequal. Two-sided confidence intervals can be used to perform approximate size 2α tests of $H_o : \sigma_A^2 = c$ against $H_a : \sigma_A^2 \neq c$.

No MVU estimator has been proposed for σ_A^2 for the case where the J_i are unequal. The unbiased ANOVA estimator is $(S_1^2 - S_2^2)/J_o$. An unbiased estimator based on S_{1U}^2 is $(S_{1U}^2 - S_2^2)/\bar{J}_H$. Both of these estimators can produce negative estimates.

Example 4.3.1 We use the data shown in Tables 4.3.1 and 4.3.2 to illustrate the formulas presented in this section. The degrees of freedom are $n_1 = 5$ and $n_2 = 29$ and the two-sided 90% confidence interval on σ_E^2 is calculated using (4.2.1) with $F_{.05:29,\infty} = 1.4675$ and $F_{.95:29,\infty} = .6106$. The computed interval is [169; 407]. To construct intervals on σ_A^2 we compute $\bar{J}_H = 6/(1/5 + \cdots + 1/9) = 4.627$, $S_{1U}^2 = 4.63[(41.20^2 + \cdots + 53.22^2) - (41.20 + \cdots + 53.22)^2/6]/5 = 610.5$, $J_o = (35 - (5^2 + \cdots + 9^2)/35)/5 = 5.669$, $m = 2$, and $M = 9$. Since S_{1U}^2 will most likely be computed with a software package, it is useful to think of the unweighted sums of squares as the sums of squares one obtains by performing an analysis of variance on the cell means. To illustrate, the following statements can be used to compute the unweighted sums of squares. This code is written using SAS©/STAT.

```
DATA BULLS;
INPUT BULL Y;
CARDS;
1 46
1 31
....
DATA GOES HERE
....
6 60
PROC SORT;
BY BULL;
*CALCULATE THE CELL MEANS AND PLACE THEM IN A NEW DATASET CALLED USS;
PROC MEANS NOPRINT;
BY BULL;
VAR Y;
OUTPUT OUT=USS MEAN=YIBAR;
*RUN THE ANOVA ON THE CELL MEANS.  THE SUM OF SQUARES FOR BULL
IS N1*S1U/JBARH;
PROC GLM DATA=USS;
CLASSES BULL;
MODEL YIBAR=BULL;
```

   Figure 1

The sums of squares that is reported for the bull effect is equal to $n_1 S_{1U}^2 / \bar{J}_H$. For this data, this value is 659.31.

Three methods can be used for constructing confidence intervals on σ_A^2. The first method is to use (4.2.2) with J_o replacing J. This method yields the 90% interval [2.85; 466]. The lower bound of this interval is consistent with the exact $\alpha = .05$ test of $H_o : \sigma_A^2 = 0$ against $H_a : \sigma_A^2 > 0$. That is, the lower bound of the 90% interval is greater than zero and H_o is rejected at the 5% level of significance. The confidence coefficient of this interval will be close to 90% if σ_A^2 is close to zero.

A second interval on σ_A^2 is obtained by replacing S_1^2 with S_{1U}^2 and J with $\bar{J}_H$ in (4.2.2). The resulting 90% interval is [0; 520]. The lower bound of this interval is zero because $S_{1U}^2 / S_2^2 = 2.46 < F_{.05:5,29} = 2.55$. Thus, in this example we are unable to reject $H_o : \sigma_A^2 = 0$ in favor of $H_a : \sigma_A^2 > 0$ when S_{1U}^2 is used as the test statistic.

The interval based on (4.3.2) is [0; 526]. The lower bound of this interval is defined to be zero because $L_m = -.291 < 0$. In comparing the three intervals on σ_A^2, it is seen that the first interval is the shortest. This is typically the case, but as shown in the simulations of Burdick and Eickman, this interval can have a confidence coefficient much less than the stated level. For this reason, it cannot generally be recommended unless it is known that σ_A^2 is close to zero. The ANOVA estimate for σ_A^2 is $(664.41 - 248.29)/5.67 = 73.4$ and the point estimate based on S_{1U}^2 is $(610.5 - 248.29)/4.63 = 78.2$.

4.3.2 Inferences on $\gamma = \sigma_A^2 + \sigma_E^2$

Burdick and Graybill (1984) recommended using (4.2.3) with S_{1U}^2 and $\bar{J}_H$ replacing S_1^2 and J, respectively, for constructing confidence intervals on γ in the unbalanced model. The resulting two-sided $1 - 2\alpha$ interval is

$$\left[\hat{\gamma} - \frac{\sqrt{G_1^2 S_{1U}^4 + G_2^2 (\bar{J}_H - 1)^2 S_2^4}}{\bar{J}_H} ; \hat{\gamma} + \frac{\sqrt{H_1^2 S_{1U}^4 + H_2^2 (\bar{J}_H - 1)^2 S_2^4}}{\bar{J}_H} \right]$$

$$(4.3.3)$$

where

$$\hat{\gamma} = \frac{S_{1U}^2 + (\bar{J}_H - 1)S_2^2}{\bar{J}_H} \qquad G_\ell = 1 - \frac{1}{F_{\alpha:n_\ell,\infty}} \quad (\ell = 1,2) \qquad \text{and}$$

$$H_\ell = \frac{1}{F_{1-\alpha:n_\ell,\infty}} - 1 \quad (\ell = 1,2)$$

Burdick and Graybill compared (4.3.3) with a Satterthwaite interval and found that (4.3.3) performed well in a variety of unbalanced designs and maintained its confidence coefficient even in situations where $\lambda_A < .25$. In contrast, the Satterthwaite procedure performed poorly and was not recommended.

Interval (4.3.3) can be used to perform an approximate 2α level test of $H_o : \gamma = c$ against $H_a : \gamma \neq c$. The null hypothesis is rejected if and only if the interval does not include c.

No MVU estimator has been proposed for γ in the unbalanced design. Two unbiased estimators are $\hat{\gamma}$ defined in (4.3.3) and the ANOVA estimator, $[S_1^2 + (J_0 - 1)S_2^2]/J_0$.

Example 4.3.2   We continue the analysis of the bull data presented in Example 4.3.1. The estimate of γ defined in (4.3.3) is $(610.5 + (3.63)(248.29))/4.63 = 327$. The computed two-sided 90% confidence interval on γ using (4.3.3) is [231; 787].

4.3.3   Inferences on Ratios of Variance Components

The problem confronted in constructing a confidence interval on either $\rho_A = \sigma_A^2/(\sigma_A^2 + \sigma_E^2)$ or $\lambda_A = \sigma_A^2/\sigma_E^2$ in the unbalanced one-fold design is the same one encountered when constructing a confidence interval on σ_A^2. Namely, the exact intervals on ρ_A and λ_A presented in Section 4.2.3 depend on the assumption that $n_1 S_1^2/\theta_1$ is an exact chi-squared random variable. Since this assumption is true in the unbalanced design if and only if $\lambda_A = 0$, these intervals cannot be used.

A possible solution to this problem is to employ the Thomas-Hultquist approximation and replace S_1^2 and J with S_{1U}^2 and $\bar{J}_H$, respectively, in the formulas of Section 4.2.3. Although this approach provides good intervals when $\lambda_A > .25$, it can be liberal in other cases. Fortunately, an exact interval on λ_A (and hence ρ_A) is available. This exact interval was proposed by Wald (1940), and is presented in Appendix B. As discussed in Appendix B, for the one-fold nested design $\min(J_1, \ldots, J_I) \leq \Delta_1 \cdots \leq \Delta_r \leq \text{Max}(J_1, \ldots, J_I)$. Additionally, $\Sigma_i^r t_i^2/\Delta_i = n_1 S_{1U}^2/\bar{J}_H$, $f = n_2$, $r = n_1$, $SSE = SS2$, and $Z(0) = S_1^2/S_2^2$. Thus, if an investigator desires a simple noniterative calculation, a conservative two-sided $1 - 2\alpha$ confidence interval on λ_A is

$$[L_m; U_M] \qquad (4.3.4)$$

where

$$L_m = \frac{S_{1U}^2}{\bar{J}_H S_2^2 F_{\alpha:n_1,n_2}} - \frac{1}{m} \qquad U_M = \frac{S_{1U}^2}{\bar{J}_H S_2^2 F_{1-\alpha:n_1,n_2}} - \frac{1}{M}$$

$$m = Min(J_1, J_2, \ldots, J_I) \qquad \text{and} \qquad M = Max(J_1, J_2, \ldots, J_I)$$

This interval was proposed by Burdick, Maqsood, and Graybill (1986). Although (4.3.4) is simple to compute, it can be appreciably wider than the exact Wald interval when the design is extremely unbalanced and λ_A is small. Thus, the exact procedure is particularly warranted if extreme unbalancedness exists. Donner, Wells, and Eliasziw (1989) have recommended the exact procedure for the type of unbalanced data encountered in family studies. Other intervals on λ_A based on normal approximations have been studied by Donner and Wells (1986) and Mian, Shoukri, and Tracy (1989) in designs where $I > 25$. Groggel, Wackerly, and Rao (1988) present a rank based method for constructing a confidence interval on a scaled version of λ_A that does not require normality. Intervals on ρ_A, ρ_E, and λ_E are obtained from intervals on λ_A by using the relationships $\rho_A = \lambda_A/(1 + \lambda_A)$, $\rho_E = 1 - \rho_A$, and $\lambda_E = 1/\lambda_A$.

Spjøtvoll (1967) derived the most powerful similar and location-invariant test for the hypothesis $H_o : \lambda_A = c$ against $H_a : \lambda_A > c$. The associated test statistic depends on the alternative value of λ_A and hence no uniformly most powerful test exits. Of course, exact hypothesis tests can be based on the Wald interval discussed in Appendix B. Spjøtvoll shows that the Wald test results from his optimal test when

Table 4.4.1   Summary Formulas for Confidence Intervals and MVU Estimators in a One-Fold Nested Design

Parameter	MVU Estimator	Confidence interval
Balanced design		
σ_E^2	S_2^2	(4.2.1)-exact
σ_A^2	$\dfrac{S_1^2 - S_2^2}{J}$	(4.2.2)-approximate
$\gamma = \sigma_A^2 + \sigma_E^2$	$\dfrac{S_1^2}{J} + S_2^2\left(1 - \dfrac{1}{J}\right)$	(4.2.3)-approximate
$*\lambda_A = \sigma_A^2/\sigma_E^2$	$\dfrac{(S_1^2/S_2^2)\left(1 - \dfrac{2}{N-1}\right) - 1}{J}$	(4.2.4)-exact
Unbalanced design		
σ_E^2	S_2^2	(4.2.1)-exact
σ_A^2	none	(4.3.2)-approximate
$\gamma = \sigma_A^2 + \sigma_E^2$	none	(4.3.3)-approximate
$*\lambda_A = \sigma_A^2/\sigma_E^2$	none	Appendix B-exact
		(4.3.4)-conservative

*Intervals on $\rho_A = \sigma_A^2/(\sigma_A^2 + \sigma_E^2)$, $\lambda_E = \sigma_E^2/\sigma_A^2$, and $\rho_E = 1 - \rho_A$ are available with appropriate transformation of the intervals on λ_A.

the alternative approaches infinity and that they are most powerful against large alternatives. These tests were also derived by Bhargava (1946). Note that if $c = 0$, Wald's test provides the usual F-test that rejects H_o when $S_1^2/S_2^2 > F_{\alpha:n_1,n_2}$. Arvesen and Layard (1975) and Prasad and Rao (1988) developed jackknife versions of Spjøtvoll's test that are robust against non-normality. Mostafa (1967) provides a locally most powerful test of $H_o: \lambda_A = c$ against $H_a: \lambda_A = c + \Delta$ where Δ is small. Additional review of inference procedures concerning λ_A and a discussion of computational considerations is provided by Verdooren (1988).

Example 4.3.3   Using the data presented in Example 4.3.1, the exact two-sided 90% interval on λ_A is computed in Appendix B. The exact upper bound in a two-sided 90% interval on λ_A is 2.16. The exact lower bound in a two-sided 90% interval on λ_A is .009. The conservative two-sided 90% interval based on (4.3.4) is [0; 2.28]. The lower bound of this interval is zero because the hypothesis $H_o : \lambda_A = 0$ cannot be rejected against $H_o : \lambda_A > 0$ using S_{1U}^2 as the test statistic. This is the same situation encountered in Example 4.3.1 for the test of σ_A^2 based on S_{1U}^2. The two-sided 90% confidence interval on ρ_A based on the exact interval for λ_A is $[.009/(1 + .009); \quad 2.16/(1 + 2.16)] = [.009; .684]$.

4.4 SUMMARY

The formulas presented for the one-fold nested design are summarized in Table 4.4.1. The unweighted sums of squares used in the unbalanced design will be generalized and used in other unbalanced designs in later chapters. All of the confidence intervals reported in this chapter are two-sided confidence intervals.

5

The Two-Fold and (Q − 1)-Fold Nested Designs

5.1 INTRODUCTION

Random models that contain only nested factors are generalizations of the one-fold nested design considered in Chapter 4. As will be illustrated in this chapter, confidence intervals for functions of variance components in nested designs can be constructed using the results of Chapters 3 and 4. Sokal and Rohlf (1969, p. 260) used a two-fold nested design to collect the data shown in Table 5.1.1. These data are from a biological experiment involving 12 female mosquito pupae that were randomly assigned into three rearing cages. The reported responses are independent measurements of the left wings of the mosquitoes.

The mosquito experiment employed a balanced two-fold nested design with four mosquitoes nested within each cage and two independent measurements taken on each left wing. All factors in the model are random. Section 5.2 presents results for the balanced two-fold nested design and Section 5.3 provides results for any balanced (Q − 1)-fold nested design. Unbalanced two-fold and (Q − 1)-fold nested designs are considered in Sections 5.4 and 5.5. Confidence

Table 5.1.1 Left Wing Measurements of Mosquitoes

Cage		1				2				3		
Mosquito	1	2	3	4	1	2	3	4	1	2	3	4
	58.5	77.8	84.0	70.1	69.8	56.0	50.7	63.8	56.6	77.8	69.9	62.1
	59.5	80.9	83.6	68.3	69.8	54.5	49.3	65.8	57.5	79.2	69.2	64.5
$\bar{Y}_{ij*}$	59.0	79.35	83.8	69.2	69.8	55.25	50.0	64.8	57.05	78.5	69.55	63.3

intervals are presented for variance components, linear combinations of variance components, and ratios of variance components.

5.2. THE BALANCED TWO-FOLD NESTED RANDOM MODEL

The balanced data set in Table 5.1.1 is represented with the balanced two-fold nested random model

$$Y_{ijk} = \mu + A_i + B_{ij} + E_{ijk}$$
$$i = 1, ..., I; \quad j = 1, ..., J; \quad k = 1,..., K \quad (5.2.1)$$

where μ is an unknown constant, A_i, B_{ij}, and E_{ijk} are mutually independent normal random variables with zero means and variances σ_A^2, σ_B^2, and σ_E^2, respectively. For the data in Table 5.1.1, $I = 3$, $J = 4$, and $K = 2$. The ANOVA table for model (5.2.1) is shown in Table 5.2.1.

Table 5.2.1 Analysis of Variance for Balanced Two-Fold Nested Random Model

SV	DF	SS[a]	MS	EMS
Factor A	$n_1 = I - 1$	SS1	S_1^2	$\theta_1 = \sigma_E^2 + K\sigma_B^2 + JK\sigma_A^2$
B within A	$n_2 = (J - 1)I$	SS2	S_2^2	$\theta_2 = \sigma_E^2 + K\sigma_B^2$
Error	$n_3 = (K - 1)IJ$	SS3	S_3^2	$\theta_3 = \sigma_E^2$
Total	$IJK - 1$	SST		

[a] $SS1 = JK\Sigma_i (\bar{Y}_{i**} - \bar{Y}_{***})^2$, $SS2 = K\Sigma_i\Sigma_j (\bar{Y}_{ij*} - \bar{Y}_{i**})^2$, $SS3 = \Sigma_i\Sigma_j\Sigma_k (Y_{ijk} - \bar{Y}_{ij*})^2$, and $SST = \Sigma_i\Sigma_j\Sigma_k (Y_{ijk} - \bar{Y}_{***})^2$.

The computed ANOVA table for the data shown in Table 5.1.1 is presented in Table 5.2.2.

Under the assumptions of model (5.2.1), $n_1 S_1^2/\theta_1$, $n_2 S_2^2/\theta_2$, and $n_3 S_3^2/\theta_3$ are jointly independent chi-squared random variables with degrees of freedom n_1, n_2, and n_3, respectively. Based on these distributional assumptions, the results of Chapter 3 are used to construct confidence intervals for parameters of interest.

5.2.1 Inferences on Variance Components

The random variable $n_3 S_3^2/\theta_3$ has a chi-squared distribution with n_3 degrees of freedom. Thus, from equation (3.2.1) an exact two-sided $1 - 2\alpha$ confidence interval on $\theta_3 = \sigma_E^2$ is

$$\left[\frac{S_3^2}{F_{\alpha:n_3,\infty}} \; ; \; \frac{S_3^2}{F_{1-\alpha:n_3,\infty}}\right] \qquad (5.2.2)$$

An exact hypothesis test concerning σ_E^2 can be based on (5.2.2) as discussed in Section 4.2.1. The MVU estimator for σ_E^2 is S_3^2.

Intervals on σ_A^2 and σ_B^2 are constructed using the results of Section 3.3.1. To see this, note from Table 5.2.1 that $\sigma_A^2 = (\theta_1 - \theta_2)/(JK)$ and $\sigma_B^2 = (\theta_2 - \theta_3)/K$. Thus, with appropriate substitution, (3.3.1) and (3.3.2) can be used for constructing approximate intervals on σ_A^2 and σ_B^2. In particular, the approximate two-sided $1 - 2\alpha$ confidence interval on σ_A^2 is

$$\left[\frac{S_1^2 - S_2^2 - \sqrt{V_L}}{JK} \; ; \; \frac{S_1^2 - S_2^2 + \sqrt{V_U}}{JK}\right] \qquad (5.2.3)$$

Table 5.2.2 Analysis of Variance for Mosquito Data

SV	DF	SS	MS
Cages	2	665.67	332.84
Mosquitoes within cages	9	1720.68	191.19
Measurement error	12	15.62	1.30
Total	23	2401.97	

where

$$V_L = G_1^2 S_1^4 + H_2^2 S_2^4 + G_{12} S_1^2 S_2^2$$

$$V_U = H_1^2 S_1^4 + G_2^2 S_2^4 + H_{12} S_1^2 S_2^2$$

$$G_\ell = 1 - \frac{1}{F_{\alpha:n_\ell,\infty}} \quad (\ell = 1,2)$$

$$H_\ell = \frac{1}{F_{1-\alpha:n_\ell,\infty}} - 1 \quad (\ell = 1,2)$$

$$G_{12} = \frac{(F_{\alpha:n_1,n_2} - 1)^2 - G_1^2 F_{\alpha:n_1,n_2}^2 - H_2^2}{F_{\alpha:n_1,n_2}} \quad \text{and}$$

$$H_{12} = \frac{(1 - F_{1-\alpha:n_1,n_2})^2 - H_1^2 F_{1-\alpha:n_1,n_2}^2 - G_2^2}{F_{1-\alpha:n_1,n_2}}$$

The approximate two-sided $1 - 2\alpha$ confidence interval on σ_B^2 is

$$\left[\frac{S_2^2 - S_3^2 - \sqrt{V_L}}{K} \; ; \; \frac{S_2^2 - S_3^2 + \sqrt{V_U}}{K} \right] \qquad (5.2.4)$$

where

$$V_L = G_2^2 S_2^4 + H_3^2 S_3^4 + G_{23} S_2^2 S_3^2$$

$$V_U = H_2^2 S_2^4 + G_3^2 S_3^4 + H_{23} S_2^2 S_3^2$$

$$G_\ell = 1 - \frac{1}{F_{\alpha:n_\ell,\infty}} \quad (\ell = 2,3)$$

$$H_\ell = \frac{1}{F_{1-\alpha:n_\ell,\infty}} - 1 \quad (\ell = 2,3)$$

$$G_{23} = \frac{(F_{\alpha:n_2,n_3} - 1)^2 - G_2^2 F_{\alpha:n_2,n_3}^2 - H_3^2}{F_{\alpha:n_2,n_3}} \quad \text{and}$$

$$H_{23} = \frac{(1 - F_{1-\alpha:n_2,n_3})^2 - H_2^2 F_{1-\alpha:n_2,n_3}^2 - G_3^2}{F_{1-\alpha:n_2,n_3}}$$

Since $\sigma_A^2 \geq 0$ and $\sigma_B^2 \geq 0$, negative bounds in (5.2.3) and (5.2.4) are defined to be zero.

Exact size α uniformly most powerful unbiased tests exist for $H_o : \sigma_A^2 = 0$ against $H_a : \sigma_A^2 > 0$ and $H_o : \sigma_B^2 = 0$ against $H_a : \sigma_B^2 > 0$ (see, e.g., Lehmann (1986, p. 424)). These tests and the MVU estimators for σ_A^2 and σ_B^2 are reported in Table 5.2.3. Table 5.2.4 reports approximate tests for other hypotheses concerning σ_A^2 and σ_B^2 that are based on (5.2.3) and (5.2.4). Hartung and Voet (1987) have proposed asymptotic chi-squared tests for these hypotheses.

Example 5.2.1   We use the ANOVA shown in Table 5.2.2 to illustrate the confidence intervals presented in this section. The variance component σ_E^2 represents error in measurement of the mosquito wings. Using (5.2.2), a two-sided 95% confidence interval on σ_E^2 is [1.30/1.9447; 1.30/.367] = [.668; 3.54]. Thus, a two-sided 95% confidence interval on the standard deviation of the measurement error, σ_E, is [.817; 1.88]. The variance component σ_B^2 is associated with variability among mosquitoes within the same cage. A two-sided 95% confidence interval on σ_B^2 is [44.6; 318]. This interval is computed using (5.2.4) with $G_2 = .5269$, $H_3 = 1.725$, $G_{23} = -.0929$, $H_2 = 2.333$, $G_3 = .4858$, and $H_{23} = -.1930$. The variance component σ_A^2 is associated with variability that arises from using different rearing cages in the experiment. Since $332.84/191.19 = 1.74 < F_{.025:2,9} = 5.71$, the lower bound on σ_A^2 is defined to be zero. In the context of hypothesis testing, we are unable to reject $H_o : \sigma_A^2 = 0$ against $H_a : \sigma_A^2 > 0$ at a

Table 5.2.3   MVU Estimators and Exact Size α Tests for σ_A^2 and σ_B^2

Parameter	MVU estimator	Hypotheses	Decision rule: Reject H_o if
σ_A^2	$\dfrac{S_1^2 - S_2^2}{JK}$	$H_o : \sigma_A^2 = 0$ $H_a : \sigma_A^2 > 0$	$\dfrac{S_1^2}{S_2^2} > F_{\alpha:n_1,n_2}$
σ_B^2	$\dfrac{S_2^2 - S_3^2}{K}$	$H_o : \sigma_B^2 = 0$ $H_a : \sigma_B^2 > 0$	$\dfrac{S_2^2}{S_3^2} > F_{\alpha:n_2,n_3}$

Table 5.2.4 Approximate Tests for σ_A^2 and σ_B^2 Based on (5.2.3) and (5.2.4)

Parameter	Hypotheses	Decision rule: Reject H_o if
σ_A^2	$H_o : \sigma_A^2 = c$ $H_a : \sigma_A^2 \neq c$	c outside of (5.2.3)
	$H_o : \sigma_A^2 \geq c$ $H_a : \sigma_A^2 < c$	Upper bound of (5.2.3) is less than c
	$H_o : \sigma_A^2 \leq c$ $H_a : \sigma_A^2 > c$	Lower bound of (5.2.3) is greater than c
σ_B^2	$H_o : \sigma_B^2 = c$ $H_a : \sigma_B^2 \neq c$	c outside of (5.2.4)
	$H_o : \sigma_B^2 \geq c$ $H_a : \sigma_B^2 < c$	Upper bound of (5.2.4) is less than c
	$H_o : \sigma_B^2 \leq c$ $H_a : \sigma_B^2 > c$	Lower bound of (5.2.4) is greater than c

significance level of .025. The upper bound on σ_A^2 computed using (5.2.3) with $\alpha = .025$ is 1616. Thus, a 95% two-sided interval on σ_A^2 is [0; 1616]. This wide interval is a consequence of having only two degrees of freedom for the cage factor. It demonstrates the lack of information contained in the data set about σ_A^2. If one is interested in this variance component, many more cages are needed in the experiment. The result that is most apparent from the analysis is that the measurement error is small relative to the variability among mosquitoes within the same cage. The MVU estimates of σ_E^2, σ_B^2, and σ_A^2 are 1.30, $(191.19 - 1.30)/2 = 94.9$, and $(332.84 - 191.19)/8 = 17.7$, respectively.

5.2.2 Inferences on Sums of Variance Components

The total variance of the response variable in model (5.2.1) is $\gamma = \sigma_A^2 + \sigma_B^2 + \sigma_E^2$. An approximate two-sided confidence interval on γ is constructed using (3.2.5) with $Q = 3$. This provides the two-sided $1 - 2\alpha$ confidence interval

$$
\left[\hat{\gamma} - \frac{\sqrt{G_1^2 S_1^4 + G_2^2(J - 1)^2 S_2^4 + G_3^2 J^2 (K - 1)^2 S_3^4}}{JK} ; \right.
$$

$$(5.2.5)$$

$$
\left. \hat{\gamma} + \frac{\sqrt{H_1^2 S_1^4 + H_2^2(J - 1)^2 S_2^4 + H_3^2 J^2 (K - 1)^2 S_3^4}}{JK} \right]
$$

where

$$
\hat{\gamma} = \frac{S_1^2 + (J - 1)S_2^2 + J(K - 1)S_3^2}{JK}
$$

$$
G_\ell = 1 - \frac{1}{F_{\alpha:n_\ell,\infty}} \quad (\ell = 1,2,3) \qquad \text{and}
$$

$$
H_\ell = \frac{1}{F_{1-\alpha:n_\ell,\infty}} - 1 \quad (\ell = 1,2,3)
$$

If one-sided intervals are desired, upper intervals (lower bounds) can be computed with (3.2.4) and lower intervals (upper bounds) with the upper bound in (5.2.5). The MVU estimator for γ is $\hat{\gamma}$.

Example 5.2.2 The two-sided 95% confidence interval on $\gamma = \sigma_A^2 + \sigma_B^2 + \sigma_E^2$ for the mosquito data is computed using (5.2.5) with $G_1 = .7289$, $G_2 = .5269$, $G_3 = .4858$, $H_1 = 38.50$, $H_2 = 2.333$, and $H_3 = 1.725$. The computed interval is [65.5; 1724] and the MVU estimate is 114. The relatively wide confidence interval on γ is the result of having $n_1 = 2$.

5.2.3 Inferences on Ratios of Variance Components

Confidence intervals on $\lambda_A = \sigma_A^2/\sigma_E^2$ are formed by writing $JK\lambda_A = (\theta_1 - \theta_2)/\theta_3$ and using the results of Wang and Graybill (1981) presented in Section 3.4.3. Since $\lambda_A \geq 0$, (3.4.5) and (3.4.6) are used to construct the approximate two-sided $1 - 2\alpha$ confidence interval on λ_A

$$\left[\frac{S_2^2}{JKS_3^2 F_{\alpha:n_1,n_3}} \left(T - F_{\alpha:n_1,\infty} + \frac{F_{\alpha:n_1,n_2}(F_{\alpha:n_1,\infty} - F_{\alpha:n_1,n_2})}{T} \right) ; \right.$$

$$\left. \frac{S_2^2}{JKS_3^2 F_{1-\alpha:n_1,n_3}} \left(T - F_{1-\alpha:n_1,\infty} + \frac{F_{1-\alpha:n_1,n_2}(F_{1-\alpha:n_1,\infty} - F_{1-\alpha:n_1,n_2})}{T} \right) \right]$$

(5.2.6)

where

$$T = \frac{S_1^2}{S_2^2}$$

If $T \leq F_{\alpha:n_1,n_2}$, the lower bound of (5.2.6) is defined to be zero and if $T \leq F_{1-\alpha:n_1,n_2}$, both bounds are zero. Approximate intervals on $\lambda_A/(1 + \lambda_A) = \sigma_A^2/(\sigma_A^2 + \sigma_E^2)$ and $1/(1 + \lambda_A) = \sigma_E^2/(\sigma_A^2 + \sigma_E^2)$ are obtained from transformations of (5.2.6). Verdooren (1988) provides an exact interval on λ_A for a given value of $\lambda_B = \sigma_B^2/\sigma_E^2$, although this quantity is almost never known. The MVU estimator of λ_A is $(S_1^2 - S_2^2)(1 - 2/n_3)/(JKS_3^2)$.

Exact intervals on λ_B, $\lambda_B/(1 + \lambda_B) = \sigma_B^2/(\sigma_B^2 + \sigma_E^2)$, and $1/(1 + \lambda_B) = \sigma_E^2/(\sigma_B^2 + \sigma_E^2)$ are the same ones used for λ_A in the one-fold nested design. As demonstrated in Section 4.2.3, these intervals are based on the exact interval for θ_2/θ_3. From Table 5.2.1 we note $\lambda_B = [(\theta_2/\theta_3) - 1]/K$ and an exact two-sided $1 - 2\alpha$ confidence interval on λ_B is

$$\left[\frac{L^* - 1}{K} ; \frac{U^* - 1}{K} \right]$$

(5.2.7)

where

$$L^* = \frac{S_2^2}{S_3^2 F_{\alpha:n_2,n_3}} \quad \text{and} \quad U^* = \frac{S_2^2}{S_3^2 F_{1-\alpha:n_2,n_3}}$$

Since $\lambda_B \geq 0$, negative bounds are defined to be zero. Exact confidence intervals on $\lambda_B/(1 + \lambda_B)$ and $1/(1 + \lambda_B)$ are obtained from

(5.2.7) using appropriate transformations. The MVU estimator for λ_B is $[(S_2^2/S_3^2)(1 - 2/n_3) - 1]/K$. Comments made in Section 4.2.3 concerning point estimation for functions of λ_A in the one-fold nested design hold for functions of λ_B in the two-fold nested design.

Approximate one-sided confidence intervals for the proportions of variability $\rho_A = \sigma_A^2/(\sigma_A^2 + \sigma_B^2 + \sigma_E^2)$ and $\rho_B = \sigma_B^2/(\sigma_A^2 + \sigma_B^2 + \sigma_E^2)$ were proposed by Graybill and Wang (1979). Based on these results, an approximate two-sided $1 - 2\alpha$ confidence interval on ρ_A is

$$\left[\frac{S_1^2 - F_{\alpha:n_1,n_2}S_2^2}{S_1^2 + (J - 1)F_{\alpha:n_1,n_2}S_2^2 + J(K - 1)F_{\alpha:n_1,n_3}S_3^2} ; \right.$$

$$\left. \frac{S_1^2 - F_{1-\alpha:n_1,n_2}S_2^2}{S_1^2 + (J - 1)F_{1-\alpha:n_1,n_2}S_2^2 + J(K - 1)F_{1-\alpha:n_1,n_3}S_3^2} \right] \tag{5.2.8}$$

The lower bound of (5.2.8) is defined to be zero if $S_1^2/S_2^2 \leq F_{\alpha:n_1,n_2}$ and both bounds are defined to be zero if $S_1^2/S_2^2 \leq F_{1-\alpha:n_1,n_2}$. The approximate two-sided $1 - 2\alpha$ confidence interval on ρ_B is

$$\left[\frac{JL_B}{1 + (J - 1)L_B} ; \frac{JU_B}{1 + (J - 1)U_B} \right] \tag{5.2.9}$$

where

$$L_B = \frac{S_2^4 - F_{\alpha:n_2,\infty}S_2^2S_3^2 - (F_{\alpha:n_2,n_3} - F_{\alpha:n_2,\infty})F_{\alpha:n_2,n_3}S_3^4}{F_{\alpha:n_2,n_1}S_1^2S_2^2 + (JK - 1)F_{\alpha:n_2,\infty}S_2^2S_3^2} \quad \text{and}$$

$$U_B = \frac{S_2^4 - F_{1-\alpha:n_2,\infty}S_2^2S_3^2 - (F_{1-\alpha:n_2,n_3} - F_{1-\alpha:n_2,\infty})F_{1-\alpha:n_2,n_3}S_3^4}{F_{1-\alpha:n_2,n_1}S_1^2S_2^2 + (JK - 1)F_{1-\alpha:n_2,\infty}S_2^2S_3^2}$$

Since $0 \leq \rho_B \leq 1$, any bound greater than one in (5.2.9) is defined to be one. In addition, if $S_2^2/S_3^2 \leq F_{\alpha:n_2,n_3}$ the lower bound is defined to be zero. If $S_2^2/S_3^2 \leq F_{1-\alpha:n_2,n_3}$, both bounds are defined to be zero. The lower bound in (5.2.9) was proposed by Graybill and Wang (1979). The upper bound results from replacing α with $1 - \alpha$ in the lower bound.

Graybill and Wang (1979) proposed another upper bound, but for practical purposes, it provides the same results as the upper bound in (5.2.9).

Graybill and Wang (1979) also proposed an approximate interval on $\rho_E = \sigma_E^2/(\sigma_A^2 + \sigma_B^2 + \sigma_E^2)$. However, a better interval is obtained using results of Section 3.4.2. In the context of model (5.2.1), define $\theta_E = [\theta_1 + (J - 1)\theta_2]/\theta_3$. Using the bounds on θ_E defined in (3.4.3) and (3.4.4) with $P = 2$ and $Q = 3$, an approximate two-sided $1 - 2\alpha$ interval on θ_E is

$$[L_E; U_E] \qquad\qquad (5.2.10)$$

where

$$L_E = \hat{\theta}_E \left[\frac{2 + k_4/(k_1k_2) - \sqrt{V_L^*}}{2(1 - k_5/k_2^2)} \right]$$

$$U_E = \hat{\theta}_E \left[\frac{2 + k_7/(k_1k_2) + \sqrt{V_U^*}}{2(1 - k_8/k_2^2)} \right]$$

$$V_L^* = (2 + k_4/(k_1k_2))^2 - 4(1 - k_5/k_2^2)(1 - k_3/k_1^2)$$

$$V_U^* = (2 + k_7/(k_1k_2))^2 - 4(1 - k_8/k_2^2)(1 - k_6/k_1^2)$$

$$k_1 = S_1^2 + (J - 1)S_2^2; \qquad k_2 = S_3^2; \qquad \hat{\theta}_E = \frac{k_1}{k_2}$$

$$k_3 = G_1^2S_1^4 + G_2^2(J - 1)^2S_2^4 + G_{12}^*(J - 1)S_1^2S_2^2,$$

$$k_4 = G_{13}S_1^2S_3^2 + G_{23}(J - 1)S_2^2S_3^2; \qquad k_5 = H_3^2S_3^4$$

$$k_6 = H_1^2S_1^4 + H_2^2(J - 1)^2S_2^4$$

$$k_7 = H_{13}S_1^2S_3^2 + H_{23}(J - 1)S_2^2S_3^2; \qquad k_8 = G_3^2S_3^4$$

$$G_{12}^* = \left(1 - \frac{1}{F_{\alpha:n_1+n_2,\infty}}\right)^2 \frac{(n_1 + n_2)^2}{n_1n_2} - \frac{G_1^2n_1}{n_2} - \frac{G_2^2n_2}{n_1}$$

$$G_{\ell 3} = \frac{(F_{\alpha:n_\ell,n_3} - 1)^2 - G_\ell^2 F_{\alpha:n_\ell,n_3}^2 - H_3^2}{F_{\alpha:n_\ell,n_3}} \qquad (\ell = 1,2)$$

$$H_{\ell 3} = \frac{(1 - F_{1-\alpha:n_\ell,n_3})^2 - H_\ell^2 F_{1-\alpha:n_\ell,n_3}^2 - G_3^2}{F_{1-\alpha:n_\ell,n_3}} \qquad (\ell = 1,2)$$

G_1, G_2, G_3, H_1, H_2, and H_3 are defined in (5.2.5). Noting that $\rho_E = [1 - 1/K + \theta_E/(JK)]^{-1}$, an approximate two-sided $1 - 2\alpha$ interval on ρ_E is

$$\left[\frac{JK}{JK - J + U_E} ; \frac{JK}{JK - J + L_E} \right] \qquad (5.2.11)$$

where L_E and U_E are defined in (5.2.10). Lu, Graybill, and Burdick (1987) proposed an alternative interval for θ_E that provides similar results.

Intervals (5.2.8), (5.2.9), and (5.2.11) can be used to compute confidence intervals on three common measures of heritability. As described by Graybill and Robertson (1957), these measures are $h_1^2 = 4\sigma_A^2/(\sigma_A^2 + \sigma_B^2 + \sigma_E^2)$, $h_2^2 = 4\sigma_B^2/(\sigma_A^2 + \sigma_B^2 + \sigma_E^2)$, and $h_3^2 = 2(\sigma_A^2 + \sigma_B^2)/(\sigma_A^2 + \sigma_B^2 + \sigma_E^2)$. By noting $h_1^2 = 4\rho_A$, $h_2^2 = 4\rho_B$, and $h_3^2 = 2(1 - \rho_E)$, one can transform either (5.2.8), (5.2.9), or (5.2.11) into an interval on the desired heritability measure.

Example 5.2.3 It was noted in Example 5.2.1 that the measurement error is small relative to the variability among mosquitoes within the same cage. It is therefore of interest to compare the relative sizes of σ_B^2 and σ_E^2 by computing the ratio $\lambda_B = \sigma_B^2/\sigma_E^2$. The MVU estimate of this ratio is $[(191.19/1.30)(1 - 2/12) - 1]/2 = 60.8$. The two-sided 95% confidence interval on λ_B is [20.9; 284]. This interval is computed using (5.2.7) with $L^* = 191.19/[(1.30)(3.4358)]$ and $U^* = 191.19/[(1.30)(.2585)]$. If we wish to express the relative magnitude of these variance components in terms of a percentage, we can compute the two-sided 95% interval on $\sigma_B^2/(\sigma_B^2 + \sigma_E^2)$, [20.9/(1 + 20.9); 284/(1 + 284)] = [.954; .996]. Based on these results, it appears the measurement error is negligible in comparison to the variation among mosquitoes, and in the future, investigators need only take one measurement on each wing.

Example 5.2.4 Graybill and Robertson (1957) report results from a study of poultry breeding. The study employed a balanced two-fold nested design in which 22 sires were each mated to 6 dams and 8 offsprings resulted from each sire-dam mating. The ANOVA table for the twelve week body weights of the offspring is shown in Table 5.2.5.

It is of interest to compute the heritability ratios $h_1^2 = 4\rho_A$, $h_2^2 = 4\rho_B$, and $h_3^2 = 2(1 - \rho_E)$. The two-sided 95% confidence interval on ρ_A is computed from (5.2.8) with $J = 6$, $K = 8$, $F_{.025:21,110} = 1.8175$, $F_{.025:21,924} = 1.7046$, $F_{.975:21,110} = .4716$, and $F_{.975:21,924} = .4874$. Placing appropriate values in (5.2.8) we obtain the two-sided 95% interval on ρ_A, [.021; .159]. By multiplying each bound by 4, the two-sided 95% confidence interval on h_1^2 is [.084; .636]. The two-sided 95% confidence interval on ρ_B is computed using (5.2.9). For these data, $F_{.975:110,\infty} = .7533$, $F_{.025:110,\infty} = 1.2811$, $F_{.025:110,924} = 1.3035$, $F_{.975:110,924} = .7435$, and the computed two-sided 95% interval on ρ_B is [.072; .192]. The corresponding 95% interval on h_2^2 is [.288; .768]. In order to compute an interval on h_3^2, we first compute the interval on θ_E shown in (5.2.10). From the data in Table 5.2.4 we compute $k_1 = 1.59$, $k_2 = .0924$, $k_3 = .118$, $k_4 = .000145$, $k_5 = .0000817$, $k_6 = .457$, $k_7 = -.00036$, $k_8 = .0000621$, $L_E = 13.24$, and $U_E = 24.76$. The corresponding two-sided 95% interval on ρ_E from (5.2.11) is [.719; .869] and the 95% confidence interval on h_3^2 is [2(1 − .869); 2(1 − .719)] = [.262; .562]. It should be noted that the interval widths are short enough to be informative in this example. This is because of the relatively large values for n_1, n_2, and n_3. This can be

Table 5.2.5 Analysis of Variance for Poultry Data

SV	DF	MS
Sires	21	.5629
Dams within sires	110	.2055
Between full siblings	924	.0924

contrasted to the uninformative intervals on σ_A^2 and γ in Examples 5.2.1 and 5.2.2 in which $n_1 = 2$.

5.2.4   Simultaneous Confidence Intervals

As discussed in Section 3.5, Kimball's inequality can be used to make joint probability statements concerning $\theta_1/\theta_3 = 1 + K\lambda_B + JK\lambda_A$ and $\theta_2/\theta_3 = 1 + K\lambda_B$. One such statement based on the lower bounds of θ_1/θ_3 and θ_2/θ_3 is

$$P[L_1 \leq 1 + K\lambda_B + JK\lambda_A \quad \text{and} \quad L_2 \leq 1 + K\lambda_B] \quad (5.2.12)$$
$$\geq (1 - \alpha_1)(1 - \alpha_2)$$

where

$$L_1 = \frac{S_1^2}{S_3^2 F_{\alpha_1:n_1,n_3}} \quad \text{and} \quad L_2 = \frac{S_2^2}{S_3^2 F_{\alpha_2:n_2,n_3}}$$

Broemeling (1969a,b) projected the confidence region in (5.2.12) onto the coordinate system defined by λ_A and λ_B to obtain the set of lower bounds

$$L_{\lambda_A} = \frac{L_1 - 1}{JK} \quad \text{and} \quad L_{\lambda_B} = \frac{\text{Max}(L_1 - 1, L_2 - 1)}{K} \quad (5.2.13)$$

In a similar manner, a set of simultaneous upper bounds on λ_A and λ_B is

$$U_{\lambda_A} = \frac{U_1 - 1}{JK} \quad U_{\lambda_B} = \frac{\text{Min}(U_1 - 1, U_2 - 1)}{K} \quad (5.2.14)$$

where

$$U_1 = \frac{S_1^2}{S_3^2 F_{1-\alpha_1:n_1,n_3}} \quad \text{and} \quad U_2 = \frac{S_2^2}{S_3^2 F_{1-\alpha_2:n_2,n_3}}$$

Since λ_A and $\lambda_B \geq 0$, negative bounds are defined to be zero. The confidence coefficients for the sets in (5.2.13) and (5.2.14) are at least as great as $(1 - \alpha_1)(1 - \alpha_2)$ and numerical results by Sahai and Anderson (1973) indicate they are very close to this value. As discussed in Section 3.5, the bounds in (5.2.13) and (5.2.14) are preferred over the simultaneous bounds proposed by Khuri (1981).

Sahai (1974) used both the Kimball and Bonferroni inequalities to construct simultaneous intervals on σ_A^2 and σ_B^2. However, as the following example illustrates, these intervals are extremely wide and shorter simultaneous intervals can be obtained using the Bonferroni inequality to combine the intervals in Section 5.2.1.

Example 5.2.5 Sahai (1974) constructed simultaneous confidence intervals on σ_A^2 and σ_B^2 using a two-fold nested data set reported in Graybill (1961, pp. 371–372). For these data, $I = 12, J = 5, K = 3$, $n_1 = 11, n_2 = 48, n_3 = 120, S_1^2 = 3.5629, S_2^2 = 1.2055,$ and $S_3^2 = .6113$. The simultaneous two-sided confidence intervals reported by Sahai are [.06; .82] for σ_A^2 and [.03; .50] for σ_B^2. The confidence coefficient for this set of two intervals is at least as great as 92%.

As an alternative approach, one could combine the intervals for σ_A^2 and σ_B^2 proposed in Section 5.2.1 using the Bonferroni inequality. If each interval has a 96% confidence coefficient, the overall confidence coefficient for the set of intervals will be at least 92%. Using equation (5.2.4) with $\alpha = .02$, we compute $F_{.02:48,120} = 1.6075, F_{.98:48,120} = .5902, G_2 = .3162, H_3 = .3279, G_{23} = .00196, H_2 = .5958, G_3 = .2204,$ and $H_{23} = -.00725$. This provides the two-sided 96% confidence interval [.05; .44] on σ_B^2. The two-sided 96% interval on σ_A^2 using (5.2.3) is [.03; .64]. Both of these intervals are shorter than the corresponding Sahai interval. We note that the Bonferroni intervals are also shorter than the ones proposed by Khuri shown in (3.5.3).

5.3 BALANCED (Q − 1)-FOLD NESTED RANDOM MODELS

The balanced $(Q - 1)$-fold nested random model is represented as

$$Y_{ij...m} = \mu + A_i + B_{ij} + \cdots + E_{ij...m} \quad (5.3.1)$$
$$i = 1, ..., I; \quad j = 1, ..., J; \quad ...; \quad m = 1, ..., M$$

where μ is an unknown constant, A_i, B_{ij}, ..., and $E_{ij...m}$ are mutually independent normal random variables with means of zero and variances σ_A^2, σ_B^2, ..., and σ_E^2, respectively. The ANOVA table for this design is shown in Table 5.3.1. For this balanced model, $n_q S_q^2/\theta_q$ has a chi-squared distribution with n_q degrees of freedom for $q = 1, ..., Q$. Furthermore, the $n_q S_q^2/\theta_q$ are jointly independent.

Using these distributional assumptions, confidence intervals on functions of variance components in model (5.3.1) can be computed using the results of Chapter 3. To illustrate, we consider the three-fold nested model

$$Y_{ijkm} = \mu + A_i + B_{ij} + C_{ijk} + E_{ijkm}$$
$$i = 1, ..., I; \quad j = 1, ..., J; \quad (5.3.2)$$
$$k = 1, ..., K; \quad m = 1, ..., M$$

where μ is an unknown constant, A_i, B_{ij}, C_{ijk}, and E_{ijkm} are mutually independent normal random variables with means of zero and variances σ_A^2, σ_B^2, σ_C^2, and σ_E^2, respectively. The ANOVA table for this model is shown in Table 5.3.2.

5.3.1 Inferences on Variance Components

In every balanced $(Q - 1)$-fold nested design, $\theta_Q = \sigma_E^2$ and $n_Q S_Q^2/\sigma_E^2$ has a chi-squared distribution with n_Q degrees of freedom. Thus, (3.2.1) will always provide an exact two-sided $1 - 2\alpha$ confidence interval on σ_E^2. All other variance components in model (5.3.1) are expressible as $\delta_q = c(\theta_q - \theta_{q+1})$ where c is a constant. For example,

Table 5.3.1 Analysis of Variance for Balanced $(Q - 1)$-Fold Nested Random Model

SV	DF	SS	MS	EMS
Factor A	n_1	$SS1$	S_1^2	θ_1
B within A	n_2	$SS2$	S_2^2	θ_2
⋮	⋮	⋮	⋮	⋮
Error	n_Q	SSQ	S_Q^2	$\theta_Q = \sigma_E^2$

Table 5.3.2 Analysis of Variance for Balanced Three-Fold Nested Random Model

SV	DF	SS	MS	EMS
Factor A	$n_1 = I - 1$	$SS1 = JKM\Sigma_i(\bar{Y}_{i***} - \bar{Y}_{****})^2$	S_1^2	$\theta_1 = \sigma_E^2 + M\sigma_C^2 + KM\sigma_B^2 + JKM\sigma_A^2$
B within A	$n_2 = I(J - 1)$	$SS2 = KM\Sigma_i\Sigma_j(\bar{Y}_{ij**} - \bar{Y}_{i***})^2$	S_2^2	$\theta_2 = \sigma_E^2 + M\sigma_C^2 + KM\sigma_B^2$
C within B within A	$n_3 = IJ(K - 1)$	$SS3 = M\Sigma_i\Sigma_j\Sigma_k(\bar{Y}_{ijk*} - \bar{Y}_{ij**})^2$	S_3^2	$\theta_3 = \sigma_E^2 + M\sigma_C^2$
Error	$n_4 = IJK(M - 1)$	$SS4 = \Sigma_i\Sigma_j\Sigma_k\Sigma_m(Y_{ijkm} - \bar{Y}_{ijk*})^2$	S_4^2	$\theta_4 = \sigma_E^2$
Total	$IJKM - 1$			

in Table 5.3.2 we note $\sigma_A^2 = (\theta_1 - \theta_2)/(JKM)$, $\sigma_B^2 = (\theta_2 - \theta_3)/(KM)$, and $\sigma_C^2 = (\theta_3 - \theta_4)/M$. Thus, confidence intervals on individual variance components can be constructed using (3.3.1) and (3.3.2).

Exact tests for $H_o : \delta_q = 0$ against $H_a : \delta_q > 0$ are provided by the statistics S_q^2/S_{q+1}^2. The decision rule is to reject H_o at the α level of significance if $S_q^2/S_{q+1}^2 > F_{\alpha:n_q,n_{q+1}}$. As an example, the test of $H_o : \sigma_A^2 = 0$ against $H_a : \sigma_A^2 > 0$ in Table 5.3.2 is to reject H_o if $S_1^2/S_2^2 > F_{\alpha:n_1,n_2}$. Approximate tests for other values of δ_q can be performed using (3.3.1) and (3.3.2) in the same manner as described in Table 5.2.4.

5.3.2   Inferences on Sums of Variance Components

The total variance of the response variable in (5.3.1) is $\gamma = \sigma_A^2 + \sigma_B^2 + \cdots + \sigma_E^2$. For constructing a confidence interval on γ, the results in Section 3.2 can be applied. For the three-fold nested design in (5.3.2), an interval on $\gamma = \sigma_A^2 + \sigma_B^2 + \sigma_C^2 + \sigma_E^2$ is formed using $Q = 4$, $c_1 = 1/(JKM)$, $c_2 = (J - 1)/(JKM)$, $c_3 = (K - 1)/(KM)$, and $c_4 = (M - 1)/M$.

If other sums of variance components are of interest, negative coefficients will be required to define the appropriate linear combinations of mean squares. For example, in Table 5.3.2 the sum $\sigma_A^2 + \sigma_C^2$ is written as $[(\theta_1 + JK\theta_3) - (\theta_2 + JK\theta_4)]/(JKM)$. When negative coefficients are required to define the linear combination of interest, the bounds in (3.3.3) and (3.3.4) can be used to construct confidence intervals.

5.3.3   Inferences on Ratios of Variance Components

Confidence intervals for several ratios of variance components can be obtained using results of Chapter 3. As an example, consider the ratio σ_C^2/σ_E^2 in Table 5.3.2. By defining $\theta_C = \theta_3/\theta_4$, it is seen $\sigma_C^2/\sigma_E^2 = (\theta_C - 1)/M$. Using (3.4.1), an exact interval $[L^*; U^*]$ can be computed for θ_C. An exact interval on σ_C^2/σ_E^2 is therefore $[(L^* - 1)/M; (U^* - 1)/M]$. Exact intervals can also be formed on $\sigma_C^2/(\sigma_C^2 + \sigma_E^2)$ and $\sigma_E^2/(\sigma_C^2 + \sigma_E^2)$ by noting $\sigma_C^2/(\sigma_C^2 + \sigma_E^2) = (\theta_C - 1)/(\theta_C - 1 + M)$ and $\sigma_E^2/(\sigma_C^2 + \sigma_E^2) = M/(\theta_C - 1 + M)$.

As another example, consider the proportion of variability $\rho_E =$

$\sigma_E^2/(\sigma_A^2 + \sigma_B^2 + \sigma_C^2 + \sigma_E^2)$ in Table 5.3.2. This ratio is a function of $\theta_E = (\theta_1 + (J - 1)\theta_2 + J(K - 1)\theta_3)/\theta_4$ by the relationship $\rho_E = [1 - 1/M + \theta_E/(JKM)]^{-1}$. A confidence interval on θ_E is constructed using (3.4.3) and (3.4.4) with $Q = 4$, $P = 3$, $c_1 = 1$, $c_2 = (J - 1)$, $c_3 = J(K - 1)$, and $c_4 = 1$. If the interval on θ_E is represented as $[L^*; U^*]$, then a confidence interval on ρ_E is

$$[(JKM)/(JKM - JK + U^*); (JKM)/(JKM - JK + L^*)]. \quad (5.3.3)$$

Example 5.3.1 Brownlee (1953, p. 117) reports the results of an industrial experiment carried out to determine whether a batch of material is homogeneous. The experiment is conducted by sampling material contained in six different vats. The material was wrung in a centrifuge and bagged. Two bags were selected at random from each vat and then two samples were taken from each bag. Finally, two independent measurements were made on each sample for the percentage of a particular ingredient. The ANOVA table for the data is shown in Table 5.3.3.

There is no evidence in the data that suggests variation exists among vats or among bags within vats. This is seen by noting $13.25/6.04 = 2.19 < F_{.05:5,6} = 4.39$ and $6.04/3.79 = 1.59 < F_{.05:6,12} = 3.00$. However, there does appear to be variation among samples within bags since $3.79/.833 = 4.55 > F_{.05:12,24} = 2.18$. Accordingly, we compute a two-sided 90% confidence interval on σ_C^2 using (3.3.1) and (3.3.2) with the subscripts 1 and 2 replaced with 3 and 4, respectively. For these data $c_3 = c_4 = 1/M$, $G_3 = .4293$, $G_4 = .3409$, $H_3 = 1.2962$, $H_4 = .7330$, $G_{34} = -.00708$, $H_{34} = -.0572$,

Table 5.3.3 Analysis of Variance for Industrial Experiment

SV	DF	MS
Vats	5	13.25
Bags within vats	6	6.04
Samples within bags	12	3.79
Analyses within samples	24	.833

and the two-sided 90% interval on σ_C^2 is [.613; 3.93]. The MVU estimate of $\sigma_C^2 = 1.48$. The total variance of a single measurement of the ingredient is $\sigma_A^2 + \sigma_B^2 + \sigma_C^2 + \sigma_E^2$. The MVU estimate of this variance is 3.775 and the two-sided 90% interval computed from (3.2.5) is [2.70; 9.84]. In order to compare the variability of samples within bags to the measurement error, we compute a confidence interval on $\sigma_C^2/(\sigma_C^2 + \sigma_E^2)$ using the transformation of (3.4.1) discussed in this section. The computed two-sided 90% interval is [.351; .839]. Based on this analysis, it cannot be concluded that the material is homogeneous.

5.4 THE UNBALANCED TWO-FOLD NESTED RANDOM MODEL

In many two-fold nested designs, it is not possible to select equal sample sizes for each level of sampling. Table 5.4.1 presents an example of this type of unbalanced design. The data shown in the table are reported by Sokal and Rohlf (1969, p. 275). The data were collected in an experiment that examined the blood pH of female mice. In the experiment 15 dams were mated over time with either two or three sires. Each sire was mated to a different dam so that 37 sires were used in the experiment. The response variable is the blood pH reading of an individual female offspring from a given dam-sire mating. The data have been coded by subtracting 7 from each response and then multiplying by 100.

The unbalanced two-fold nested random model is represented as

$$Y_{ijk} = \mu + A_i + B_{ij} + E_{ijk} \qquad (5.4.1)$$
$$i = 1, \dots, I; \qquad j = 1, \dots, J_i; \qquad k = 1, \dots, K_{ij}$$

where μ, A_i, B_{ij}, and E_{ijk} are defined in (5.2.1). For example, the data shown in Table 5.4.1 has $I = 15$, $J_1 = 2$, $J_3 = 3$, $K_{41} = 3$, $K_{51} = 5$, etc. Table 5.4.2 reports a commonly used analysis of variance for model (5.4.1).

The sums of squares shown in Table 5.4.2 are referred to as the Type I sums of squares where $\bar{Y}_{ij*} = \Sigma_k Y_{ijk}/K_{ij}$, $\bar{Y}_{i**} = \Sigma_j\Sigma_k Y_{ijk}/K_{i*}$, $\bar{Y}_{***} = \Sigma_i\Sigma_j\Sigma_k Y_{ijk}/K_{**}$, $K_{i*} = \Sigma_j K_{ij}$, and $K_{**} = \Sigma_i\Sigma_j K_{ij}$. The coefficients in the EMS column are defined as $w_1 = \Sigma_i\Sigma_j[K_{ij}^2(1/K_{i*} - 1/K_{**})]/n_1$, $w_2 = \Sigma_i K_{i*}^2(1/K_{i*} - 1/K_{**})/n_1$, and $w_3 = \Sigma_i\Sigma_j K_{ij}^2(1/K_{ij}$

Table 5.4.1 Blood pH Readings of Mice

Dam number	Sire number	Blood pH readings of individual mice				
1	1	48	48	52	54	
	2	48	53	43	39	
2	1	45	43	49	40	40
	2	50	45	43	36	
3	1	40	45	42	48	
	2	45	33	40	46	
	3	40	47	40	47	47
4	1	38	48	46		
	2	37	31	45	41	
5	1	44	51	49	51	52
	2	49	49	49	50	
	3	48	59	59		
6	1	54	36	36	40	
	2	44	47	48	48	
	3	43	52	50	46	39
7	1	41	42	36	47	
	2	47	36	43	38	41
	3	53	40	44	40	45
8	1	52	53	48		
	2	40	48	50	40	51
9	1	40	34	37	45	
	2	42	37	46	40	
10	1	39	31	30	41	48
	2	50	44	40	45	
11	1	52	54	52	56	53
	2	56	39	52	49	48
12	1	50	45	43	44	49
	2	52	43	38	33	
13	1	39	37	33	43	42
	2	43	38	44		
	3	46	44	37	54	
14	1	50	53	51	43	
	2	44	45	39	52	
	3	42	48	45	51	48
15	1	47	49	45	43	42
	2	45	42	52	51	32
	3	51	51	53	45	51

Table 5.4.2 Analysis of Variance with Type I SS for Unbalanced
Two-Fold Nested Random Model

SV	DF	SS[a]	MS	EMS
Factor A	$n_1 = I - 1$	SS1	S_1^2	$\theta_1 = \sigma_E^2 + w_1\sigma_B^2 + w_2\sigma_A^2$
B within A	$n_2 = \Sigma_i J_i - I$	SS2	S_2^2	$\theta_2 = \sigma_E^2 + w_3\sigma_B^2$
Error	$n_3 = \Sigma_i\Sigma_j K_{ij} - \Sigma_i J_i$	SS3	S_3^2	$\theta_3 = \sigma_E^2$
Total	$\Sigma_i\Sigma_j K_{ij} - 1$	SST		

[a]$SS1 = \Sigma_i\Sigma_j K_{ij}(\bar{Y}_{i**} - \bar{Y}_{***})^2$, $SS2 = \Sigma_i\Sigma_j K_{ij}(\bar{Y}_{ij*} - \bar{Y}_{i**})^2$, $SS3 = \Sigma_i\Sigma_j\Sigma_k(Y_{ijk} - \bar{Y}_{ij*})^2$, and $SST = \Sigma_i\Sigma_j\Sigma_k(Y_{ijk} - \bar{Y}_{***})^2$.

$- 1/K_{i*})/n_2$. Although $n_3 S_3^2/\theta_3$ is a chi-squared random variable with n_3 degrees of freedom, neither $n_1 S_1^2/\theta_1$ nor $n_2 S_2^2/\theta_2$ are in general chi-squared random variables. Additionally, S_1^2 and S_2^2 are in general not independent. Cummings and Gaylor (1974) report two exceptions to these general results. In particular, if $K_{ij} = K_i$, S_1^2 and S_2^2 are independent. If $J_i = 2$, $K_{i1} = K_1$, and $K_{i2} = K_2$, then both $n_1 S_1^2/\theta_1$ and $n_2 S_2^2/\theta_2$ have chi-squared distributions. Finally, if $K_{ij} = K$, then S_1^2, S_2^2, and S_3^2 are jointly independent and $n_2 S_2^2/\theta_2$ has a chi-squared distribution with n_2 degrees of freedom (see, e.g., Burdick and Graybill (1985) and Scheffé (1959, p. 252)). The mean square S_3^2 is independent of both S_1^2 and S_2^2 for all values of J_i and K_{ij}.

In a balanced design where $J_i = J$ and $K_{ij} = K$, Table 5.4.2 reduces to Table 5.2.1. Thus, $w_2 = JK$ and $w_1 = w_3 = K$. In addition, $n_1 S_1^2/\theta_1$, $n_2 S_2^2/\theta_2$, and $n_3 S_3^2/\theta_3$ are jointly independent chi-squared random variables with n_1, n_2, and n_3 degrees of freedom, respectively.

5.4.1 Inferences on Variance Components

Since $n_3 S_3^2/\theta_3$ is a chi-squared random variable in the unbalanced design, interval (5.2.2) provides an exact $1 - 2\alpha$ confidence interval on σ_E^2. The MVU estimator for σ_E^2 is S_3^2.

The strategy employed for constructing intervals on the other variance components is the same one used in Chapter 4. Specifically, unweighted sums of squares are used in the balanced model equations presented in Chapter 3. Although the assumption that mean squares are jointly independent chi-squared random variables is not generally satisfied, this approach works well in most commonly encountered types of unbalancedness. Confidence intervals on σ_A^2 and σ_B^2 were obtained in this manner by Hernandez, Burdick, and Birch (1992). They employed the unweighted mean squares

$$S_{1U}^2 = w_{2U} \sum_i (\bar{Y}_{i*}^* - \bar{Y}_*^*)^2/n_1, \quad \text{and} \quad (5.4.2)$$

$$S_{2U}^2 = w_{3U} \sum_i \sum_j (\bar{Y}_{ij*} - \bar{Y}_{i*}^*)^2/n_2$$

where $\bar{Y}_{ij*} = \Sigma_k Y_{ijk}/K_{ij}$, $\bar{Y}_{i*}^* = \Sigma_j \bar{Y}_{ij*}/J_i$, $\bar{Y}_*^* = \Sigma_i \bar{Y}_{i*}^*/I$, $w_{2U} = I/\Sigma_i 1/J_i\bar{K}_{Hi}$, $w_{3U} = n_2/\Sigma_i(J_i - 1)/(\bar{K}_{Hi})$, and $\bar{K}_{Hi} = J_i/(\Sigma_j 1/K_{ij})$ represents the harmonic mean of the K_{ij} values in level i of factor A. Table 5.4.3 summarizes this approach in the form of an ANOVA table where $w_{1U} = [\Sigma_i 1/J_i]/[\Sigma_i 1/(J_i\bar{K}_{Hi})]$. We note that when $J_i = J$ and $K_{ij} = K$, then $S_{1U}^2 = S_1^2$, $\theta_{1U} = \theta_1$, $S_{2U}^2 = S_2^2$, $\theta_{2U} = \theta_2$, $w_{1U} = w_1 = K$, $w_{2U} = w_2 = JK$, and $w_{3U} = w_3 = K$.

In many experimental situations an investigator can select equal subsamples in the final stage of a nested design. Tietjen (1974) calls this condition "last stage uniformity." In this situation, model (5.4.1) has $K_{ij} = K$, $\bar{K}_{Hi} = K$, $w_{1U} = w_{3U} = K$, $w_{2U} = IK/\Sigma_i 1/J_i$, and $S_{2U}^2 = S_2^2$. As previously noted, $n_2 S_2^2/\theta_2 = n_2 S_{2U}^2/\theta_{2U}$ has an exact chi-squared distribution with n_2 degrees of freedom in this case. Another simplification that occurs when $K_{ij} = K$ is that both $n_1 S_1^2/\theta_1$ and $n_1 S_{1U}^2/\theta_{1U}$ are independent of $n_2 S_2^2/\theta_2$.

Table 5.4.3 Analysis of Variance with Unweighted SS for Unbalanced Two-Fold Nested Random Model

SV		DF	SS[a]	MS	EMS
Factor A	$n_1 = I - 1$		$SS1U$	S_{1U}^2	$\theta_{1U} = \sigma_E^2 + w_{1U}\sigma_B^2 + w_{2U}\sigma_A^2$
B within A	$n_2 = \Sigma_i J_i - I$		$SS2U$	S_{2U}^2	$\theta_{2U} = \sigma_E^2 + w_{3U}\sigma_B^2$
Error	$n_3 = \Sigma_i\Sigma_j K_{ij} - \Sigma_i J_i$		$SS3$	S_3^2	$\theta_3 = \sigma_E^2$

[a] $SS1U = w_{2U}\Sigma_i(\bar{Y}_{i*}^* - \bar{Y}_*^*)^2$, $SS2U = w_{3U}\Sigma_i\Sigma_j(\bar{Y}_{ij*} - \bar{Y}_{i*}^*)^2$, $SS3 = \Sigma_i\Sigma_j\Sigma_k(Y_{ijk} - \bar{Y}_{ij*})^2$

In the case where $J_i \neq J$ and $K_{ij} \neq K$, neither $n_1 S_{1U}^2/\theta_{1U}$ nor $n_2 S_{2U}^2/\theta_{2U}$ in general have chi-squared distributions. Furthermore, they are not independent. However, we treat these mean squares as if they are jointly independent chi-squared random variables and apply the results of Section 3 to construct the desired intervals. By writing $\sigma_B^2 = (\theta_{2U} - \theta_3)/w_{3U}$, and $\sigma_A^2 = [\theta_{1U} - (w_{1U}/w_{3U})\theta_{2U} + (w_{1U}/w_{3U} - 1)\theta_3]/w_{2U}$, equations (3.3.3) and (3.3.4) can be used to construct confidence intervals on σ_A^2 and σ_B^2. Hernandez et al. (1992) show that the resulting intervals generally maintain the stated confidence coefficient over a wide range of conditions. The recommended approximate two-sided $1 - 2\alpha$ confidence interval on σ_B^2 is

$$\left[\frac{S_{2U}^2 - S_3^2 - \sqrt{V_L}}{w_{3U}} ; \frac{S_{2U}^2 - S_3^2 + \sqrt{V_U}}{w_{3U}} \right] \tag{5.4.3}$$

where

$$V_L = G_2^2 S_{2U}^4 + H_3^2 S_3^4 + G_{23} S_{2U}^2 S_3^2$$

$$V_U = H_2^2 S_{2U}^4 + G_3^2 S_3^4 + H_{23} S_{2U}^2 S_3^2$$

and G_2, G_3, H_2, H_3, G_{23}, and H_{23} are defined in (5.2.4). When $K_{ij} = K$, interval (5.4.3) simplifies to (5.2.4). This interval performs well over a variety of conditions. It can be liberal, however, when σ_B^2/σ_E^2 is small, and the design is extremely unbalanced. In these situations, Hernandez et al. (1992) recommended the approximate two-sided $1 - 2\alpha$ interval on σ_B^2

$$\left[\frac{S_{2U}^2 L_m}{F_{\alpha:n_2,\infty}(1 + w_{3U}L_m)} ; \frac{S_{2U}^2 U_M}{F_{1-\alpha:n_2,\infty}(1 + w_{3U}U_M)} \right] \tag{5.4.4}$$

where

$$L_m = \frac{S_{2U}^2}{F_{\alpha:n_2,n_3} w_{3U} S_3^2} - \frac{1}{m}$$

$$U_M = \frac{S_{2U}^2}{F_{1-\alpha:n_2,n_3}w_3 U S_3^2} - \frac{1}{M}$$

$$m = \text{Min}(K_{ij}), \quad \text{and} \quad M = \text{Max}(K_{ij})$$

If $L_m < 0$, the lower bound in (5.4.4) is defined to be zero. Similarly, if $U_M < 0$, the upper bound is defined to be zero. The interval in (5.4.4) is an extension of (4.3.2) and is generally conservative. Interval (5.4.4) simplifies to the Tukey-Williams interval discussed in Section 3.3.1 when $K_{ij} = K$.

Hernandez et al. (1992) recommended using the unweighted mean squares in (3.3.3) and (3.3.4) for constructing a confidence interval on σ_A^2. This approximate two-sided $1 - 2\alpha$ confidence interval on σ_A^2 is

$$\left[\frac{S_{1U}^2 - c_2 S_{2U}^2 + c_3 S_3^2 - \sqrt{V_L}}{w_{2U}} ; \frac{S_{1U}^2 - c_2 S_{2U}^2 + c_3 S_3^2 + \sqrt{V_U}}{w_{2U}} \right]$$

$$(5.4.5)$$

where

$$\begin{aligned}
V_L &= G_1^2 S_{1U}^4 + G_3^2 c_3^2 S_3^4 + H_2^2 c_2^2 S_{2U}^4 + G_{12} c_2 S_{1U}^2 S_{2U}^2 \\
&\quad + G_{32} c_3 c_2 S_3^2 S_{2U}^2 + G_{13}^* c_3 S_{1U}^2 S_3^2 \qquad\qquad \text{if } c_3 \geq 0 \\
&= G_1^2 S_{1U}^4 + H_2^2 c_2^2 S_{2U}^4 + H_3^2 c_3^2 S_3^4 \\
&\quad + G_{12} c_2 S_{1U}^2 S_{2U}^2 + G_{13} |c_3| S_{1U}^2 S_3^2 \qquad\qquad \text{if } c_3 < 0 \\
V_U &= H_1^2 S_{1U}^4 + H_3^2 c_3^2 S_3^4 + G_2^2 c_2^2 S_{2U}^4 + H_{12} c_2 S_{1U}^2 S_{2U}^2 \\
&\quad + H_{32} c_3 c_2 S_3^2 S_{2U}^2 \qquad\qquad\qquad\qquad\quad \text{if } c_3 \geq 0 \\
&= H_1^2 S_{1U}^4 + G_2^2 c_2^2 S_{2U}^4 + G_3^2 c_3^2 S_3^4 \\
&\quad + H_{12} c_2 S_{1U}^2 S_{2U}^2 + H_{13} |c_3| S_{1U}^2 S_3^2 + H_{23}^* c_2 |c_3| S_{2U}^2 S_3^2 \quad \text{if } c_3 < 0
\end{aligned}$$

$$c_2 = \frac{w_{1U}}{w_{3U}}; \qquad c_3 = c_2 - 1$$

$$G_\ell = 1 - \frac{1}{F_{\alpha:n_\ell,\infty}} \quad (\ell = 1,2,3)$$

$$H_\ell = \frac{1}{F_{1-\alpha:n_\ell,\infty}} - 1 \quad (\ell = 1,2,3)$$

$$G_{1\ell} = \frac{(F_{\alpha:n_1,n_\ell} - 1)^2 - G_1^2 F_{\alpha:n_1,n_\ell}^2 - H_\ell^2}{F_{\alpha:n_1,n_\ell}} \quad (\ell = 2,3)$$

$$G_{32} = \frac{(F_{\alpha:n_3,n_2} - 1)^2 - G_3^2 F_{\alpha:n_3,n_2}^2 - H_2^2}{F_{\alpha:n_3,n_2}}$$

$$G_{13}^* = \left(1 - \frac{1}{F_{\alpha:n_1+n_3,\infty}}\right)^2 \frac{(n_1 + n_3)^2}{n_1 n_3} - \frac{G_1^2 n_1}{n_3} - \frac{G_3^2 n_3}{n_1}$$

$$H_{1\ell} = \frac{(1 - F_{1-\alpha:n_1,n_\ell})^2 - H_1^2 F_{1-\alpha:n_1,n_\ell}^2 - G_\ell^2}{F_{1-\alpha:n_1,n_\ell}} \quad (\ell = 2,3)$$

$$H_{32} = \frac{(1 - F_{1-\alpha:n_3,n_2})^2 - H_3^2 F_{1-\alpha:n_3,n_2}^2 - G_2^2}{F_{1-\alpha:n_3,n_2}} \qquad \text{and}$$

$$H_{23}^* = \left(1 - \frac{1}{F_{\alpha:n_2+n_3,\infty}}\right)^2 \frac{(n_2 + n_3)^2}{n_2 n_3} - \frac{G_2^2 n_2}{n_3} - \frac{G_3^2 n_3}{n_2}$$

These intervals perform well over a wide variety of unbalanced designs. Additionally, Burdick, Birch, and Graybill (1986) provide simulation results that demonstrate interval (5.4.5) performs well when $K_{ij} = K$. An exception occurs if $\rho_A = \sigma_A^2/(\sigma_A^2 + \sigma_B^2 + \sigma_E^2)$ is small and the J_i are extremely unequal with $J_i = 1$ or $J_i = 2$ for a few of the I groups. In these situations the confidence intervals can be liberal. If

$J_i = J$ and $K_{ij} = K$, then $w_{1U} = w_{3U} = K$, $w_{2U} = JK$, $c_2 = 1$, $c_3 = 0$, $S_{1U}^2 = S_1^2$, $S_{2U}^2 = S_2^2$, and (5.4.5) reduces to (5.2.3).

Hernandez et al. (1992) also examined the performance of intervals based on S_1^2 and S_2^2. They found them to be slightly inferior to the unweighted sums of squares intervals (5.4.3)–(5.4.5).

We have previously noted simplifications that result if $K_{ij} = K$ or if $J_i = J$ and $K_{ij} = K$. One other special case is worth noting. If $J_i = J$, so that only the last stage of the design is unbalanced, a method proposed by Khuri (1990) can be used to construct a set of jointly independent sums of squares that have exact chi-squared distributions. The associated mean squares, S_{1K}^2, S_{2K}^2, and S_{3K}^2, have degrees of freedom n_1, n_2, and $K_{**} - 2IJ + 1 > 0$, respectively, with expected values $E(S_{1K}^2) = J\sigma_A^2 + \sigma_B^2 + \lambda_{max}\sigma_E^2 = \theta_{1K}$, $E(S_{2K}^2) = \sigma_B^2 + \lambda_{max}\sigma_E^2 = \theta_{2K}$, and $E(S_{3K}^2) = \sigma_E^2 = \theta_{3K}$ where λ_{max} is the maximum eigenvalue of a particular matrix. If $K_{ij} = K$, then $\lambda_{max} = 1/K$, $KS_{1K}^2 = S_1^2$, and $KS_{2K}^2 = S_2^2$. Since $\sigma_A^2 = (\theta_{1K} - \theta_{2K})/J$, and $\sigma_B^2 = \theta_{2K} - \lambda_{max}\theta_{3K}$, one can use (3.3.1) and (3.3.2) to construct intervals on σ_A^2 and σ_B^2. The major disadvantage of this approach, however, is that in order to compute S_{1K}^2, S_{2K}^2, and S_{3K}^2, one must perform a non-unique partitioning of the error sums of squares, $SS3$. Alternative partitionings result in different values for S_{1K}^2, S_{2K}^2, and S_{3K}^2, and different computed confidence intervals. Thus, two independent investigators could use the same data set and come to entirely different conclusions. A second disadvantage of Khuri's method is that the computations are complex, and no generally available software program exists for performing the computations. Finally, Khuri's sums of squares can produce wider intervals than the unweighted sums of squares because there are only $K_{**} - 2IJ + 1$ degrees of freedom associated with σ_E^2. Given these disadvantages, we recommend (5.4.3)–(5.4.5) for constructing intervals on σ_A^2 and σ_B^2 even in the special case where $J_i = J$.

The test for $H_o : \sigma_B^2 = 0$ against $H_a : \sigma_B^2 > 0$ shown in Table 5.2.3 is still exact in the unbalanced model. This is because $n_2 S_2^2/\theta_2$ is independent of $n_3 S_3^2/\theta_3$ and has a chi-squared distribution under the null hypothesis. This test corresponds to the Wald test discussed in Section 5.4.3. Hussein and Milliken (1978b) provide an exact test of $H_o : \sigma_B^2 = 0$ when Var $(B_{ij}) = d_{ij}\sigma_B^2$ where d_{ij} are known constants.

The test for $H_o : \sigma_A^2 = 0$ against $H_a : \sigma_A^2 > 0$ shown in Table 5.2.3 is not exact under model (5.4.1). This is because in the unbalanced design neither $n_1 S_1^2/\theta_1$ nor $n_2 S_2^2/\theta_2$ are in general chi-squared random variables under the null hypothesis $H_o : \sigma_A^2 = 0$. Tietjen (1974) recommended that one still use the test shown in Table 5.2.3 and noted that it is exact for the special case where $K_{ij} = K$. For this special case $n_2 S_2^2/\theta_2$ has an exact chi-squared distribution and $n_1 S_1^2/\theta_1$ has an exact chi-squared distribution under the null hypothesis. Cummings and Gaylor (1974) recommended two alternative tests based on the Type I sums of squares and the Satterthwaite approximation. Their results suggest that non-chi-squaredness and dependence have canceling effects on the size of the test. Thus, they concluded such an approach will perform similarly to a Satterthwaite approximation in a balanced design. Tan and Cheng (1984) compared these three tests with a fourth test statistic based on a Satterthwaite approximation involving the Type I sums of squares. Their results suggest that the test in Table 5.2.3 cannot be recommended for extremely unbalanced designs, but that the other three tests can generally be recommended. Khuri (1987) proposed an exact test for $H_o : \sigma_A^2 = 0$ and compared the power of his test to the four tests examined by Tan and Cheng. When $J_i = J$, Khuri's test statistic reduces to S_{1K}^2/S_{2K}^2 where S_{1K}^2 and S_{2K}^2 are the means squares defined earlier in this section. Although Khuri's exact test appears to have greater power than the approximate tests recommended by Tan and Cheng, it has the disadvantage that computation of the test statistic requires a non-unique partitioning of the error sums of squares. Given this disadvantage, we recommend testing $H_o : \sigma_A^2 = 0$ with the lower bound defined in (5.4.5). In particular, one rejects $H_o : \sigma_A^2 = 0$ against $H_a : \sigma_A^2 > 0$ if the lower bound in (5.4.5) is greater than zero. Hernandez et al. (1992) showed this test has comparable power to the approximate tests studied by Tan and Cheng and is better at maintaining the stated test size. The power of the test is also comparable to Khuri's exact test. Hussein and Milliken (1978b) provide an exact test of $H_o : \sigma_A^2 = 0$ where Var$(A_i) = d_i\sigma_A^2$, Var(B_{ij}) $= \sigma_B^2$, and $K_{ij} = K$. If $d_i = d$, the test reduces to the one reported in Table 5.2.3.

Example 5.4.1 We use the data shown in Table 5.4.1 to illustrate the formulas presented in this section. Using the definitions in Table 5.4.3 and equation (5.4.2), we compute $n_1 = 14$, $n_2 = 22$, $n_3 = 123$, $S_2^2 = 36.37$, $S_{1U}^2 = 128.6$, $S_{2U}^2 = 36.64$, $S_3^2 = 24.74$, $w_{1U} = 4.186$, $w_{2U} = 9.914$, and $w_{3U} = 4.222$. The unweighted mean squares are most easily obtained by computing sums of squares based on cell means. To illustrate, the following SAS© code performs these calculations.

```
DATA DAMS;
INPUT DAM SIRE Y;
CARDS;
1 1 48
....
DATA GOES HERE
....
15 3 51
PROC SORT; BY DAM SIRE;
*CALCULATE CELL MEANS FOR EACH SIRE;
PROC MEANS NOPRINT;BY DAM SIRE;
VAR Y;
OUTPUT OUT=USS1  MEAN=YIJSTR;
*CALCULATE CELL MEANS FOR EACH DAM;
PROC MEANS NOPRINT;BY DAM;
VAR YIJSTR;
OUTPUT OUT=USS2  MEAN=YISTR;
*RUN ANOVA ON SIRE CELL MEANS.  THE SUM OF SQUARES FOR
SIRE IS N2*S2U/W3U;
PROC GLM DATA=USS1;
CLASSES DAM SIRE;
MODEL YIJSTR= DAM SIRE(DAM);
*RUN ANOVA ON DAM CELL MEANS.  THE SUM OF SQUARES FOR DAM
IS N1*S1U/W2U;
PROC GLM DATA=USS2;
CLASSES DAM;
MODEL YISTR=DAM;
```

The sire effect sum of squares for data set $USS1$ corresponds to $n_2 S_{2U}^2 / w_{3U}$ and is equal to 190.932. The dam effect sum of squares for $USS2$ corresponds to $n_1 S_{1U}^2 / w_{2U}$ and is equal to 181.608.

The variance component σ_B^2 corresponds to the variation among sires within dams. We employ the exact test shown in Table 5.2.3 to test $H_o : \sigma_B^2 = 0$ against $H_a : \sigma_B^2 > 0$. Since $S_2^2 / S_3^2 = 36.37/24.74 = 1.47 < F_{.05;22,123} = 1.63$, there is no evidence to suggest variation exists among sires. The two-sided 90% confidence interval on σ_A^2 using (5.4.5) is [3.31; 23.9]. In this calculation $c_3 < 0$, $G_1 = .4089$, $G_2 = .3515$, $G_3 = .1794$, $H_1 = 1.131$, $H_2 = .7831$, $H_3 = .2502$, $G_{12} = -.0126$, $G_{13} = .005259$, $H_{12} = -.0434$, $H_{13} = -.0213$, and $H_{23}^* = .0151$. Since the lower bound on σ_A^2 is greater than zero, we reject $H_o : \sigma_A^2 = 0$ in favor of $H_a : \sigma_A^2 > 0$. We conclude that variability in blood pH of females can in part be explained by variability in mothers.

5.4.2   Inferences on Sums of Variance Components

Intervals on $\gamma = \sigma_A^2 + \sigma_B^2 + \sigma_E^2$ can be constructed using unweighted sums of squares by writing $\gamma = [\theta_{1U} + c_{2U}\theta_{2U} + c_{3U}\theta_3]/w_{2U}$ where $c_{2U} = (w_{2U} - w_{1U})/w_{3U} \geq 0$, and $c_{3U} = w_{2U} - 1 - c_{2U} \geq 0$. If one ignores the dependence and non-chi-squaredness of the unweighted mean squares, (3.2.5) can be used to construct a two-sided interval on γ. The resulting approximate $1 - 2\alpha$ interval on γ is

$$\left[\frac{S_{1U}^2 + c_{2U}S_{2U}^2 + c_{3U}S_3^2 - \sqrt{V_L}}{w_{2U}} ; \frac{S_{1U}^2 + c_{2U}S_{2U}^2 + c_{3U}S_3^2 + \sqrt{V_U}}{w_{2U}} \right]$$

(5.4.6)

where

$$V_L = G_1^2 S_{1U}^4 + G_2^2 c_{2U}^2 S_{2U}^4 + G_3^2 c_{3U}^2 S_3^4$$
$$V_U = H_1^2 S_{1U}^4 + H_2^2 c_{2U}^2 S_{2U}^4 + H_3^2 c_{3U}^2 S_3^4$$

$$G_\ell = 1 - \frac{1}{F_{\alpha:n_\ell,\infty}} \quad (\ell = 1,2,3) \quad \text{and}$$

$$H_\ell = \frac{1}{F_{1-\alpha:n_\ell,\infty}} - 1 \quad (\ell = 1,2,3)$$

Equation (3.2.4) with appropriate substitutions can be used for forming a lower bound on γ. Hernandez and Burdick (1992) provide simulations that illustrate (5.4.6) generally maintains the stated confidence coefficient. Burdick and Graybill (1985) provide additional simulations for the special case where $K_{ij} = K$.

Example 5.4.2 Continuing the analysis of Table 5.4.1 we construct a two-sided 90% confidence interval on γ using (5.4.6). Using the calculations shown in Example 5.4.1, $c_{2U} = 1.357$, and $c_{3U} = 7.5575$. The computed interval on γ is [30.31; 52.74].

5.4.3 Inferences on Ratios of Variance Components

Confidence intervals on $\rho_A = \sigma_A^2/\gamma$ and $\rho_B = \sigma_B^2/\gamma$ were developed by Sen, Graybill, and Ting (1992). They employed S_{1U}^2, S_{2U}^2, and S_3^2 in a modification of (5.2.8) and (5.2.9). The resulting approximate $1 - \alpha$ lower bound on ρ_A is

$$\frac{w_{3U}S_{1U}^2 - w_{1U}F_{\alpha:n_1,n_2}S_{2U}^2 - (w_{3U} - w_{1U})F_{\alpha:n_1,n_3}S_3^2}{w_{3U}S_{1U}^2 - (w_{1U} - w_{2U})F_{\alpha:n_1,n_2}S_{2U}^2 - (w_{3U} - w_{1U} + w_{2U} - w_{2U}w_{3U})F_{\alpha:n_1,n_3}S_3^2}$$

$$(5.4.7)$$

The lower bound is defined to be zero if $(w_{3U}S_{1U}^2)/(w_{1U}S_{2U}^2) < F_{\alpha:n_1,n_2}$. The approximate upper bound is obtained from (5.4.7) by replacing α with $1 - \alpha$. The approximate lower bound on ρ_B is

$$\frac{w_{2U}L_B}{w_{3U} + (w_{2U} - w_{1U})L_B} \qquad (5.4.8)$$

where

$$L_B = \frac{S_{2U}^4 - F_{\alpha:n_2,\infty}S_{2U}^2 S_3^2 - (F_{\alpha:n_2,n_3} - F_{\alpha:n_2,\infty})F_{\alpha:n_2,n_3}S_3^4}{F_{\alpha:n_2,n_1}S_{1U}^2 S_{2U}^2 + (w_{2U} - 1)F_{\alpha:n_2,\infty}S_{2U}^2 S_3^2}$$

If $S_{2U}^2/S_3^2 \le F_{\alpha:n_2,n_3}$, the lower bound is defined to be zero. An approximate $1 - \alpha$ upper bound on ρ_B is computed using (5.4.8) and replacing α with $1 - \alpha$.

Sen et al. (1992) also modified the degrees of freedom associated with S_{1U}^2 and S_{2U}^2 using the Satterthwaite approximation. However, unless extreme unbalancedness exists, this modification is not needed. When $K_{ij} = K$ and $J_i = J$, (5.4.7) and (5.4.8) reduce to (5.2.8) and (5.2.9), respectively. Burdick, Birch, and Graybill (1986) provide additional results for the special case where $K_{ij} = K$.

An exact interval on $\lambda_B = \sigma_B^2/\sigma_E^2$ can be computed using Wald's method as described in Appendix B. A numerical example is given in Example 5.4.3. A conservative interval on λ_B is obtained from (B.7) by noting that $\min(K_{ij}) \le \Delta_1 \le \cdots \le \Delta_r \le \max(K_{ij})$, $\Sigma_i t_i^2/\Delta_i = n_2 S_{2U}^2/w_{3U}$, $f = n_3$, $r = n_2$, and $SSE = SS3$. The resulting conservative $1 - 2\alpha$ interval on λ_B is

$$[L_m ; U_M] \tag{5.4.9}$$

where L_m and U_M are defined in (5.4.4). This interval simplifies to (5.2.7) when $K_{ij} = K$.

An exact interval on $\lambda_A = \sigma_A^2/\sigma_E^2$ cannot be computed using Wald's method (see, e.g., Seely and El-Bassiouni, (1983)). An exact confidence interval on λ_A for a given value of λ_B is provided by Verdooren (1988).

Example 5.4.3 The 95% lower bound on ρ_A for the data in Table 5.4.1 can be computed using (5.4.7). For these data the lower bound is .087. The 95% upper bound is .481. Since $S_{2U}^2/S_3^2 < F_{.05:22,123,} = 1.6295$, the 95% lower bound on ρ_B is zero.

The exact Wald interval on λ_B is computed as described in Appendix B. Since $S_2^2/S_3^2 = 1.47 < F_{.05:22,123}$, the lower bound on λ_B is zero. For illustrative purposes we compute the upper bound on λ_B. Following the approach outlined in Appendix B, we treat μ and A_i as fixed effects and B_{ij} as a random effect. The fixed effects design matrix for the data in Table 5.4.1 is

$$
X = \begin{bmatrix} 1 & 1 & 0 & \cdots & 0 \\ 1 & 0 & 1 & \cdots & 0 \\ \vdots & \vdots & \vdots & \ddots & \vdots \\ 1 & 0 & 0 & \cdots & 1 \end{bmatrix} \begin{matrix} \} K_{1*} = 8 \text{ rows} \\ \} K_{2*} = 9 \text{ rows} \\ \vdots \\ \} K_{15*} = 15 \text{ rows} \end{matrix}
$$

The matrix X has 160 rows and 16 columns. The Z matrix used to represent the random nested effect is

$$
Z = \begin{bmatrix} 1 & 0 & \cdots & 0 \\ 0 & 1 & \cdots & 0 \\ \vdots & \vdots & \ddots & \vdots \\ 0 & 0 & \cdots & 1 \end{bmatrix} \begin{matrix} \} K_{11} = 4 \text{ rows} \\ \} K_{12} = 4 \text{ rows} \\ \vdots \\ \} K_{15,3} = 5 \text{ rows} \end{matrix}
$$

This matrix is formed by multiplying columns that represent A and B main effects in the same manner that an interaction effect is formed in a crossed model. The matrix Z has 160 rows and 45 columns. Some columns, e.g., column 3, contain all zeros. This is because not all dams are mated with 3 sires. In computing the 95% upper bound on λ_B we obtain $r = 22, f = 123, p^* = 15, \Delta_1 = 3.333$, and $\Delta_{22} = 5.0$. The exact 95% upper bound based on 10 iterations of the bisection algorithm is .403. The conservative 95% upper bound based on (5.4.9) is .444.

5.5 UNBALANCED (Q − 1)-FOLD NESTED DESIGNS

In unbalanced (Q − 1)-fold nested designs, we employ the same approach demonstrated in Section 5.4. In particular, we recommend using unweighted sums of squares in the Chapter 3 equations. Although these sums of squares are in general neither chi-squared nor jointly independent, they perform satisfactorily and provide better intervals than those based on Type I sums of squares. To demonstrate our approach, consider the unbalanced three-fold ($Q = 4$) model

$$
\begin{aligned}
Y_{ijkm} &= \mu + A_i + B_{ij} + C_{ijk} + E_{ijkm} \\
i &= 1, ..., I; \qquad j = 1, ..., J_i; \\
k &= 1, ..., K_{ij}; \qquad m = 1, ..., M_{ijk}
\end{aligned}
\tag{5.5.1}
$$

where μ, A_i, B_{ij}, C_{ijk}, and E_{ijkm} are mutually independent normal random variables with means of zero and variances σ_A^2, σ_B^2, σ_C^2, and σ_E^2, respectively. Table 5.5.1 shows an ANOVA table for model (5.5.1) using unweighted sums of squares. In defining the unweighted sums squares $\bar{Y}_{ijk*} = \Sigma_m Y_{ijkm}/M_{ijk}$, $\bar{Y}_{ij*}^* = \Sigma_k \bar{Y}_{ijk*}/K_{ij}$, $\bar{Y}_{i*}^* = \Sigma_j \bar{Y}_{ij*}^* /J_i$, and $\bar{Y}_*^* = \Sigma_i \bar{Y}_{i*}^* /I$. The constants w_{1U}^*, ..., w_{6U}^* are defined as

$$
w_{1U}^* = \frac{w_{3U}^*}{w_{2U}} \qquad w_{2U}^* = \frac{w_{1U} w_{3U}^*}{w_{2U}} \qquad w_{3U}^* = \frac{I}{\Sigma_i \left(\Sigma_j \frac{1}{\bar{M}_{Hij} K_{ij}} \right) / J_i^2}
$$

$$
w_{4U}^* = \frac{w_{5U}^*}{w_{3U}} \qquad w_{5U}^* = \frac{n_2}{\Sigma_i \dfrac{J_i - 1}{J_i} \left(\Sigma_j \dfrac{1}{\bar{M}_{Hij} K_{ij}} \right)}
\tag{5.5.2}
$$

$$
w_{6U}^* = \frac{n_3}{\Sigma_i \Sigma_j \dfrac{K_{ij} - 1}{\bar{M}_{Hij}}} \qquad \bar{M}_{Hij} = \frac{K_{ij}}{\Sigma_k \dfrac{1}{M_{ijk}}}
$$

and w_{1U}, w_{2U}, and w_{3U} are defined after equation (5.4.2). Given the unweighted expected mean squares in Table 5.5.1, the appropriate linear combinations needed to define the variance components can be determined. Equations (3.3.3) and (3.3.4) can then be used to construct confidence intervals on the variance components.

Wald's procedure can be used to obtain an exact interval on $\sigma_{Q-1}^2/\sigma_E^2$ where σ_{Q-1}^2 represents the variance component associated with the $(Q - 1)$ stage of nesting. If all stages except the last stage are balanced, the results of Khuri (1990) can be used to form a set of Q jointly independent sums of squares that have chi-squared distribu-

Table 5.5.1 Analysis of Variance with Unweighted SS for Unbalanced Three-Fold Nested Random Model

SV	DF	SS	MS	EMS
Factor A	$n_1 = I - 1$	$SS1U = w_{3U}^* \sum_i (\bar{Y}_{i*}^* - \bar{Y}_{**}^*)^2$	S_{1U}^2	$\theta_{1U} = \sigma_E^2 + w_{1U}^* \sigma_C^2 + w_{2U}^* \sigma_B^2 + w_{3U}^* \sigma_A^2$
B within A	$n_2 = \sum_i J_i - 1$	$SS2U = w_{5U}^* \sum_i \sum_j (\bar{Y}_{ij*}^* - \bar{Y}_{i*}^*)^2$	S_{2U}^2	$\theta_{2U} = \sigma_E^2 + w_{4U}^* \sigma_C^2 + w_{5U}^* \sigma_B^2$
C within B	$n_3 = \sum_i \sum_j K_{ij} - \sum_i J_i$	$SS3U = w_{6U}^* \sum_i \sum_j \sum_k (\bar{Y}_{ijk*} - \bar{Y}_{ij*}^*)^2$	S_{3U}^2	$\theta_{3U} = \sigma_E^2 + w_{6U}^* \sigma_C^2$
Error	$n_4 = \sum_i \sum_j \sum_k M_{ijk} - \sum_i \sum_j K_{ij}$	$SS4 = \sum_i \sum_j \sum_k \sum_m (Y_{ijkm} - \bar{Y}_{ijk*})^2$	S_4^2	$\theta_4 = \sigma_E^2$

Table 5.5.2 Insecticide Residue on Celery

Plot	Sample	Subsample	Residue on celery	
1	1	1	.52	.43
		2	.40	.52
	2	1	.26	
		2	.54	
	3	1	.52	
2	1	1	.50	.59
		2	.47	.50
	2	1	.04	
		2	.43	
	3	1	1.08	
3	1	1	.34	.26
		2	.32	.45
	2	1	.25	
		2	.38	
	3	1	.29	
4	1	1	.18	.24
		2	.31	.29
	2	1	.13	
		2	.25	
	3	1	.10	
5	1	1	1.05	.66
		2	.60	.51
	2	1	.95	
		2	.84	
	3	1	.92	
6	1	1	.52	.66
		2	.55	.40
	2	1	.33	
		2	.26	
	3	1	.41	
7	1	1	.77	.56
		2	.51	.60
	2	1	.44	
		2	.50	
	3	1	.44	

Table 5.5.2   (Continued)

Plot	Sample	Subsample	Residue on celery	
8	1	1	.89	.92
		2	.75	.58
	2	1	.64	
		2	.54	
	3	1	.36	
9	1	1	.50	.67
		2	.60	.53
	2	1	.60	
		2	.71	
	3	1	.92	
10	1	1	.58	.52
		2	.56	.44
	2	1	.46	
		2	.52	
	3	1	.52	
11	1	1	.24	.36
		2	.48	.30
	2	1	.53	
		2	.50	
	3	1	.39	

tions. Although these sums of squares can be used in the formulas of Chapter 3, we do not recommend this approach since the derived sums of squares are not unique.

Example 5.5.1   Table 5.5.2 reports data from Bliss (1967, p. 354) concerning measurements of insecticide residue on celery. Insecticide was sprayed on plots of celery and then residue was measured from plants selected in a three-stage nested design. The experimental design can be represented with model (5.5.1) where $I = 11$, $J_i = 3$ for all i, $K_{ij} = 1$ or 2, and $M_{ijk} = 1$ or 2. Table 5.5.3 displays the unweighted sums of squares for the data in Table 5.5.2. These sums of squares can be computed by performing an analysis of variance on the cell means as demonstrated with the two-fold design in Example 5.4.1. To illustrate, the following SAS© statements will compute the unweighted sum of squares for the subsamples within samples.

Table 5.5.3 Analysis of Variance for Insecticide
Experiment Using Unweighted Sums of Squares

SV	DF	SS
Plots	10	$SS1U = 1.707$
Samples within plots	22	$SS2U = 1.089$
Subsamples within samples	22	$SS3U = .3414$
Determinations	22	$SS4 = .2209$

```
DATA CELERY;
INPUT PLOT SAMPLE SUB Y;
CARDS;
1 1 1 .52
....
DATA GOES HERE
....
11 3 1 .39
PROC SORT;
BY PLOT SAMPLE SUB;
PROC MEANS NOPRINT;
BY PLOT SAMPLE SUB;
VAR Y;
OUTPUT OUT=USS1 MEAN=YIJKSTR;
PROC GLM DATA=USS1;
CLASSES PLOT SAMPLE SUB;
MODEL YIJKSTR=PLOT SAMPLE(PLOT) SUB(SAMPLE PLOT);
```

The sum of squares for SUB(SAMPLE PLOT) in USS1 is equal to $n_3 S_{3U}^2/w_{6U}^* = .2560375$. A confidence interval on σ_B^2 is formed by writing $\sigma_B^2 = [\theta_{2U} - (w_{4U}^*/w_{6U}^*)\theta_{3U} + (w_{4U}^*/w_{6U}^* - 1)\theta_4]/w_{5U}^*$ and using (3.3.3) and (3.3.4) with appropriate substitutions. Using the sums of squares in Table 5.5.3, a two-sided 95% confidence interval on σ_B^2 is [.0065; .0494]. This interval is computed using $G_2 = G_3 = G_4 = .4019$, $H_2 = H_3 = H_4 = 1.003$, $G_{23} = G_{24} = H_{23} = H_{24} = -.0256$, $H_{34}^* = .0731$, $w_{4U}^* = 1.143$, $w_{5U}^* = 1.714$, and $w_{6U}^* = 1.333$. An approximate test for $H_o : \sigma_B^2 = 0$ against $H_a : \sigma_B^2 > 0$ is to reject H_o if the lower bound of the confidence interval is greater than zero. Using $\alpha = .025$ we conclude that there exists significant

Table 5.6.1 Summary Formulas for Confidence Intervals and MVU Estimators in the Two-Fold Balanced Nested Design

Parameter	MVU estimator	Confidence interval (approximate unless noted)
σ_E^2	S_3^2	(5.2.2)-Exact
σ_A^2	$\dfrac{S_1^2 - S_2^2}{JK}$	(5.2.3)
σ_B^2	$\dfrac{S_2^2 - S_3^2}{K}$	(5.2.4)
$\gamma = \sigma_A^2 + \sigma_B^2 + \sigma_E^2$	$\dfrac{S_1^2 + (J - 1)S_2^2 + J(K - 1)S_3^2}{JK}$	(5.2.5)
$\lambda_A = \dfrac{\sigma_A^2}{\sigma_E^2}$	$\dfrac{(S_1^2 - S_2^2)(1 - 2/n_3)}{JKS_3^2}$	(5.2.6) (5.2.13), (5.2.14) Simultaneous
$\lambda_B = \dfrac{\sigma_B^2}{\sigma_E^2}$	$\dfrac{(S_2^2/S_3^2)(1 - 2/n_3) - 1}{K}$	(5.2.7)-Exact (5.2.13), (5.2.14) Simultaneous
$\rho_A = \dfrac{\sigma_A^2}{\gamma}$	None	(5.2.8)
$\rho_B = \dfrac{\sigma_B^2}{\gamma}$	None	(5.2.9)
$\rho_E = \dfrac{\sigma_E^2}{\gamma}$	None	(5.2.11)

Table 5.6.2 Summary Formulas for Confidence Intervals in the
$(Q - 1)$-Fold Balanced Nested Design

General form	Three-fold parameter	Confidence interval (approximate unless noted)
θ_Q	σ_E^2	(3.2.1)-Exact
$c(\theta_q - \theta_{q+1})$	$\sigma_A^2,\ \sigma_B^2,\ \sigma_C^2$	(3.3.1), (3.3.2)
$\Sigma_q^Q c_q \theta_q$	$\sigma_A^2 + \sigma_B^2 + \sigma_C^2 + \sigma_E^2$	(3.2.5)-Two-sided (3.2.4)-Lower bound (3.2.5)-Upper bound
$\dfrac{\theta_{Q-1}}{\theta_Q}$	$\dfrac{\sigma_C^2}{\sigma_E^2},\ \dfrac{\sigma_C^2}{\sigma_C^2 + \sigma_E^2},\ \dfrac{\sigma_E^2}{\sigma_C^2 + \sigma_E^2}$	(3.4.1)-Exact
$\dfrac{c(\theta_q - \theta_{q+1})}{\theta_Q}$	$\dfrac{\sigma_A^2}{\sigma_E^2},\ \dfrac{\sigma_B^2}{\sigma_E^2},\ \dfrac{\sigma_A^2}{\sigma_A^2 + \sigma_E^2}$ $\dfrac{\sigma_E^2}{\sigma_A^2 + \sigma_E^2},\ \dfrac{\sigma_B^2}{\sigma_B^2 + \sigma_E^2},\ \dfrac{\sigma_E^2}{\sigma_B^2 + \sigma_E^2}$	(3.4.5), (3.4.6)
$\dfrac{\sum_{q=1}^{Q-1} c_q \theta_q}{\theta_Q}$	$\dfrac{\sigma_E^2}{\sigma_A^2 + \sigma_B^2 + \sigma_C^2 + \sigma_E^2}$	(3.4.3), (3.4.4)

variation among the samples within each plot. Confidence intervals on σ_A^2 and σ_C^2 can be constructed in a similar manner.

5.6 SUMMARY

The confidence intervals presented in this chapter for the two-fold balanced nested design are summarized in Table 5.6.1. Most of these results are based on the formulas presented in Chapter 3. Table 5.6.2 summarizes general forms that are used in the $(Q - 1)$-fold balanced nested design and illustrates their application in a three-fold balanced nested design. Modifications for unbalanced designs were discussed in Sections 5.4 and 5.5.

6

Crossed Random Designs

6.1 INTRODUCTION

This chapter considers random models in which all factors are crossed. A two-factor crossed model was introduced in Example 2.2.2 where the quality control manager of a company designed an experiment to measure the sources of variability in the manufacturing of window screens. The data collected in the experiment are displayed in Table 6.1.1. The data are balanced because an equal number of observations are obtained for each of the 12 operator-machine combinations. Since there is more than one replication of each combination, the interaction effect between machines and operators can be estimated. The balanced two-factor crossed design with interaction is presented in Section 6.2.

Other designs that involve crossed factors are examined in this chapter. Section 6.3 considers the balanced two-factor crossed design with no interaction, Section 6.4 considers the balanced three-factor crossed design with interaction, and Sections 6.5–6.7 consider several unbalanced factorial designs. As in previous chapters, confidence intervals are provided for individual variance components, sums of variance components, and ratios of variance components. Intervals on

Table 6.1.1   Screen Lengths in Inches from Quality
Control Experiment

Machine	Operator		
	1	2	3
1	36.3	35.2	35.8
	35.2	36.3	34.9
2	36.7	35.3	36.0
	36.1	36.2	36.1
3	35.1	36.8	35.9
	35.2	36.1	36.1
4	35.2	34.9	36.3
	36.8	34.9	36.1

variance components and sums of variance components for crossed
designs are based on results of Chapter 3. Since we have illustrated the
application of these general formulas in Chapters 4 and 5, only brief
results are provided in this chapter. Detailed results are presented for
some ratio forms not previously considered.

6.2   THE BALANCED TWO-FACTOR CROSSED RANDOM MODEL WITH INTERACTION

The data in Table 6.1.1 are modeled as a balanced two-factor crossed
random model with interaction. Notationally, this model is written as

$$Y_{ijk} = \mu + A_i + B_j + (AB)_{ij} + E_{ijk}$$
$$i = 1, ..., I; \qquad j = 1, ..., J; \qquad k = 1, ..., K > 1 \quad (6.2.1)$$

where μ is an unknown constant, A_i, B_j, $(AB)_{ij}$, and E_{ijk} are mutually
independent normal random variables with means of zero and vari-
ances σ_A^2, σ_B^2, σ_{AB}^2, and σ_E^2, respectively. For the data reported in Table
6.1.1, $I = 4$, $J = 3$, and $K = 2$. The ANOVA table for model (6.2.1)
is shown in Table 6.2.1.

Under the assumptions of model (6.2.1), $n_q S_q^2 / \theta_q$ are jointly in-

Table 6.2.1 Analysis of Variance for Balanced Two-Factor Crossed Random Model with Interaction

SV	DF	SS	MS	EMS
A	$n_1 = I - 1$	$SS1 = JK\Sigma_i(\bar{Y}_{i**} - \bar{Y}_{***})^2$	S_1^2	$\theta_1 = \sigma_E^2 + K\sigma_{AB}^2 + JK\sigma_A^2$
B	$n_2 = J - 1$	$SS2 = IK\Sigma_j(\bar{Y}_{*j*} - \bar{Y}_{***})^2$	S_2^2	$\theta_2 = \sigma_E^2 + K\sigma_{AB}^2 + IK\sigma_B^2$
AB	$n_3 = (I - 1)(J - 1)$	$SS3 = K\Sigma_i\Sigma_j(\bar{Y}_{ij*} - \bar{Y}_{i**} - \bar{Y}_{*j*} + \bar{Y}_{***})^2$	S_3^2	$\theta_3 = \sigma_E^2 + K\sigma_{AB}^2$
Error	$n_4 = IJ(K - 1)$	$SS4 = \Sigma_i\Sigma_j\Sigma_k(Y_{ijk} - \bar{Y}_{ij*})^2$	S_4^2	$\theta_4 = \sigma_E^2$
Total	$IJK - 1$	$SST = \Sigma_i\Sigma_j\Sigma_k(Y_{ijk} - \bar{Y}_{***})^2$		

dependent chi-squared random variables with n_q degrees of freedom, respectively, for $q = 1, \ldots, 4$. These distributional assumptions allow the use of Chapter 3 results to construct confidence intervals on several important functions of variance components.

6.2.1 Inferences on Variance Components

Since $\theta_4 = \sigma_E^2$, equation (3.2.1) provides the exact $1 - 2\alpha$ confidence interval on σ_E^2. Approximate confidence intervals on the other variance components can be constructed using the results of Section 3.3.1. In particular, (3.3.1) and (3.3.2) can be used to construct intervals on $\sigma_A^2 = (\theta_1 - \theta_3)/(JK)$, $\sigma_B^2 = (\theta_2 - \theta_3)/(IK)$, and $\sigma_{AB}^2 = (\theta_3 - \theta_4)/K$. To illustrate, the approximate $1 - 2\alpha$ confidence interval on σ_A^2 is

$$\left[\frac{S_1^2 - S_3^2 - \sqrt{V_L}}{JK} ; \frac{S_1^2 - S_3^2 + \sqrt{V_U}}{JK} \right] \qquad (6.2.2)$$

where

$$V_L = G_1^2 S_1^4 + H_3^2 S_3^4 + G_{13} S_1^2 S_3^2$$

$$V_U = H_1^2 S_1^4 + G_3^2 S_3^4 + H_{13} S_1^2 S_3^2$$

$$G_\ell = 1 - \frac{1}{F_{\alpha:n_\ell,\infty}} \qquad (\ell = 1, 3)$$

$$H_\ell = \frac{1}{F_{1-\alpha:n_\ell,\infty}} - 1 \qquad (\ell = 1, 3)$$

$$G_{13} = \frac{(F_{\alpha:n_1,n_3} - 1)^2 - G_1^2 F_{\alpha:n_1,n_3}^2 - H_3^2}{F_{\alpha:n_1,n_3}} \qquad \text{and}$$

$$H_{13} = \frac{(1 - F_{1-\alpha:n_1,n_3})^2 - H_1^2 F_{1-\alpha:n_1,n_3}^2 - G_3^2}{F_{1-\alpha:n_1,n_3}}$$

Confidence intervals on σ_B^2 and σ_{AB}^2 are similarly defined with the appropriate substitutions.

Although no exact intervals exist for σ_A^2, σ_B^2, and σ_{AB}^2, exact tests for $H_o : \sigma_A^2 = 0$, $H_o : \sigma_B^2 = 0$ and $H_o : \sigma_{AB}^2 = 0$ are available. These tests and the MVU estimators are reported in Table 6.2.2.

Example 6.2.1. The ANOVA table shown in Table 6.2.3 was reported by Weir (1949). The table is the outcome of an experiment that examined strain and daily differences in blood pH of five inbred strains of mice. The five strains used in the study were viewed as a sample from all available strains. Likewise, the time variable was viewed as a random factor because interest was in the magnitude of fluctuation in blood pH over the study period and not in comparison of average daily levels. Thus, the random model (6.2.1) is appropriate for modeling the

Table 6.2.2 MVU Estimators and Exact Size α Tests for the Variance Components in Model (6.2.1)

Parameter	MVU Estimator	Hypotheses	Decision Rule: Reject H_o if
σ_A^2	$\dfrac{S_1^2 - S_3^2}{JK}$	$H_o : \sigma_A^2 = 0$ $H_a : \sigma_A^2 > 0$	$\dfrac{S_1^2}{S_3^2} > F_{\alpha:n_1,n_3}$
σ_B^2	$\dfrac{S_2^2 - S_3^2}{IK}$	$H_o : \sigma_B^2 = 0$ $H_a : \sigma_B^2 > 0$	$\dfrac{S_2^2}{S_3^2} > F_{\alpha:n_2,n_3}$
σ_{AB}^2	$\dfrac{S_3^2 - S_4^2}{K}$	$H_o : \sigma_{AB}^2 = 0$ $H_a : \sigma_{AB}^2 > 0$	$\dfrac{S_3^2}{S_4^2} > F_{\alpha:n_3,n_4}$

Table 6.2.3 ANOVA Table for Mice Data

SV	DF	MS
Mouse strains	4	0.0920
Days of test	5	0.0101
Interaction	20	0.0052
Error	120	0.0034

data. For this experiment $I = 5$ strains, $J = 6$ days, and $K = 5$ mice. A primary objective of the study was to determine the differences in blood pH among strains of mice. To examine these differences a confidence interval on σ_A^2 is desired. A 95% confidence on σ_A^2 using (6.2.2) is [.0009; .0251]. This interval is computed using $\alpha = .025$, $G_1 = .6410$, $G_3 = .4147$, $H_1 = 7.257$, $H_3 = 1.085$, $G_{13} = .0198$, and $H_{13} = -.9482$. Based on this interval, it appears there exists a genetic strain effect that accounts for variability in blood pH.

6.2.2 Inferences on Sums of Variance Components

The total variation of the response variable in model (6.2.1) is $\gamma = \sigma_A^2 + \sigma_B^2 + \sigma_{AB}^2 + \sigma_E^2$. Confidence intervals on γ can be constructed using the results of Section 3.2. For example, a two-sided interval on γ can be constructed using (3.2.5) with $Q = 4$, $\hat{\gamma} = c_1 S_1^2 + c_2 S_2^2 + c_3 S_3^2 + c_4 S_4^2$, $c_1 = 1/(JK)$, $c_2 = 1/(IK)$, $c_3 = (IJ - I - J)/(IJK)$, and $c_4 = (K - 1)/K$. The MVU estimator of γ is $\hat{\gamma}$.

In order to define other sums of variance components, one may require negative coefficients on some of the expected mean squares. For example, the total variance due to the controllable factors is $\sigma_A^2 + \sigma_B^2 + \sigma_{AB}^2 = (I\theta_1 + J\theta_2 + (IJ - I - J)\theta_3 - IJ\theta_4)/(IJK)$ and θ_4 has a negative coefficient. In such a situation, one should use the bounds shown in (3.3.3) and (3.3.4).

6.2.3 Inferences on Ratios of Variance Components

Intervals on $\lambda_A = \sigma_A^2/\sigma_E^2 = (\theta_1 - \theta_3)/(JK\theta_4)$ and $\lambda_B = \sigma_B^2/\sigma_E^2 = (\theta_2 - \theta_3)/(IK\theta_4)$ are constructed using the bounds in (3.4.5) and (3.4.6). To illustrate, an approximate $1 - 2\alpha$ confidence interval on λ_A is $[L ; U]$ where

$$L = \frac{S_3^2}{JKS_4^2 \, F_{\alpha:n_1,n_4}} \left(T - F_{\alpha:n_1,\infty} + \frac{F_{\alpha:n_1,n_3}(F_{\alpha:n_1,\infty} - F_{\alpha:n_1,n_3})}{T} \right)$$

$$U = \frac{S_3^2}{JKS_4^2 \, F_{1-\alpha:n_1,n_4}} \left(T - F_{1-\alpha:n_1,\infty} + \frac{F_{1-\alpha:n_1,n_3}(F_{1-\alpha:n_1,\infty} - F_{1-\alpha:n_1,n_3})}{T} \right)$$

$$(6.2.3)$$

where

$$T = \frac{S_1^2}{S_3^2}$$

If $T \le F_{\alpha:n_1,n_3}$, the lower bound in (6.2.3) is defined to be zero and if $T \le F_{1-\alpha:n_1,n_3}$, both bounds are defined to be zero.

Since $\lambda_{AB} = \sigma_{AB}^2/\sigma_E^2 = (\theta_3/\theta_4 - 1)/K$, exact confidence intervals are constructed on λ_{AB} using (3.4.1). After appropriate transformation, an exact $1 - 2\alpha$ confidence interval on λ_{AB} is

$$\left[\frac{L^* - 1}{K} ; \frac{U^* - 1}{K} \right] \tag{6.2.4}$$

where

$$L^* = \frac{S_3^2}{S_4^2 F_{\alpha:n_3,n_4}} \quad \text{and} \quad U^* = \frac{S_3^2}{S_4^2 F_{1-\alpha:n_3,n_4}}$$

Exact confidence intervals are formed on $\sigma_{AB}^2/(\sigma_{AB}^2 + \sigma_E^2)$ and $\sigma_E^2/(\sigma_{AB}^2 + \sigma_E^2)$ from (6.2.4) by noting $\sigma_{AB}^2/(\sigma_{AB}^2 + \sigma_E^2) = \lambda_{AB}/(1 + \lambda_{AB})$ and $\sigma_E^2/(\sigma_{AB}^2 + \sigma_E^2) = 1/(1 + \lambda_{AB})$. Simultaneous intervals on λ_A, λ_B, and λ_{AB} can be derived from the results of Broemeling (1969a,b) shown in (3.5.1).

A ratio of interest that is in a form not previously considered in earlier chapters is $\lambda_A/\lambda_B = \sigma_A^2/\sigma_B^2 = [I(\theta_1 - \theta_3)]/[J(\theta_2 - \theta_3)]$. This ratio is useful in comparing the variability of the two main effect factors. The distinguishing feature of the ratio of expected mean squares is that θ_3 appears with a negative sign in both the numerator and the denominator. Thus, the general results of Section 3.4.2 are not appropriate for this problem. Ting and Graybill (1991) considered the general problem of setting a confidence interval on $(\theta_1 - \theta_3)/(\theta_2 - \theta_3)$. Using these results, an approximate $1 - 2\alpha$ confidence interval on σ_A^2/σ_B^2 is

$$\left[\frac{I(S_1^2 - F_{\alpha:n_1,n_3}S_3^2)}{JF_{\alpha:n_1,n_2}(S_2^2 - F_{1-\alpha:n_2,n_3}S_3^2)} ; \frac{I(S_1^2 - F_{1-\alpha:n_1,n_3}S_3^2)}{JF_{1-\alpha:n_1,n_2}(S_2^2 - F_{\alpha:n_2,n_3}S_3^2)} \right] \tag{6.2.5}$$

Notice that for both bounds in (6.2.5) to make sense, it is necessary for $S_1^2/S_3^2 > F_{\alpha:n_1,n_3}$ and $S_2^2/S_3^2 > F_{\alpha:n_2,n_3}$. These restrictions do not limit the usefulness of (6.2.5) since the ratio σ_A^2/σ_B^2 will only be of interest if evidence suggests that both σ_A^2 and σ_B^2 are positive. In this case, the required evidence is that both $H_o : \sigma_A^2 = 0$ and $H_o : \sigma_B^2 = 0$ be rejected at a level of α. For instance, in Example 6.2.1 there is not sufficient evidence to conclude $\sigma_B^2 > 0$ and hence an interval on σ_A^2/σ_B^2 is not of interest. Interval (6.2.5) is exact when either (i) $\theta_3 = 0$, or (ii) $I \to \infty$ with J fixed, or (iii) $J \to \infty$ with I fixed. Simulation results presented by Ting and Graybill indicate the true confidence coefficient is generally above the stated level.

The proportions of variability $\rho_A = \sigma_A^2/\gamma$, $\rho_B = \sigma_B^2/\gamma$, and $\rho_{AB} = \sigma_{AB}^2/\gamma$, where $\gamma = \sigma_A^2 + \sigma_B^2 + \sigma_{AB}^2 + \sigma_E^2$ are other important ratios that cannot be written in the general form presented in Section 3.4.2. For example, $\rho_A = I(\theta_1 - \theta_3)/[I\theta_1 + J\theta_2 + (IJ - I - J)\theta_3 + IJ(K - 1)\theta_4]$ and θ_3 appears in both the numerator and denominator of the ratio. Leiva and Graybill (1986) considered this special form and derived approximate intervals for all three proportions of variability. Their approximate $1 - 2\alpha$ confidence interval on ρ_A is

$$\left[\frac{IL_A}{IL_A + J} \quad ; \quad \frac{IU_A}{IU_A + J} \right] \qquad (6.2.6)$$

where

$$L_A = \frac{S_1^2 - F_{\alpha:n_1,n_3}S_3^2}{I(K - 1)F_{\alpha:n_1,\infty} S_4^2 + F_{\alpha:n_1,n_2}S_2^2 + (I - 1)F_{\alpha:n_1,\infty} S_3^2} \qquad \text{and}$$

$$U_A = \frac{S_1^2 - F_{1-\alpha:n_1,n_3}S_3^2}{I(K - 1)F_{1-\alpha:n_1,\infty} S_4^2 + F_{1-\alpha:n_1,n_2}S_2^2 + (I - 1)F_{1-\alpha:n_1,\infty} S_3^2}.$$

An approximate $1 - 2\alpha$ confidence interval on ρ_B is

$$\left[\frac{JL_B}{JL_B + I} \quad ; \quad \frac{JU_B}{JU_B + I} \right] \qquad (6.2.7)$$

where

$$L_B = \frac{S_2^2 - F_{\alpha:n_2,n_3}S_3^2}{J(K-1)F_{\alpha:n_2,\infty}S_4^2 + F_{\alpha:n_2,n_1}S_1^2 + (J-1)F_{\alpha:n_2,\infty}S_3^2} \quad \text{and}$$

$$U_B = \frac{S_2^2 - F_{1-\alpha:n_2,n_3}S_3^2}{J(K-1)F_{1-\alpha:n_2,\infty}S_4^2 + F_{1-\alpha:n_2,n_1}S_1^2 + (J-1)F_{1-\alpha:n_2,\infty}S_3^2}.$$

An approximate $1-2\alpha$ confidence interval on ρ_{AB} is

$$\left[\frac{IJL_{AB}}{L_{AB}(IJ-I-J)+IJ} \quad ; \quad \frac{IJU_{AB}}{U_{AB}(IJ-I-J)+IJ} \right] \quad (6.2.8)$$

where

$$L_{AB} = \frac{IJ(S_3^2 - F_{\alpha:n_3,n_4}S_4^2)}{IF_{\alpha:n_3,n_1}S_1^2 + JF_{\alpha:n_3,n_2}S_2^2 + (IJK-I-J)F_{\alpha:n_3,n_4}S_4^2} \quad \text{and}$$

$$U_{AB} = \frac{IJ(S_3^2 - F_{1-\alpha:n_3,n_4}S_4^2)}{IF_{1-\alpha:n_3,n_1}S_1^2 + JF_{1-\alpha:n_3,n_2}S_2^2 + (IJK-I-J)F_{1-\alpha:n_3,n_4}S_4^2}.$$

All negative bounds in (6.2.6)–(6.2.8) are defined to be zero.

The simulation results reported by Leiva and Graybill indicate intervals (6.2.6)–(6.2.8) provide confidence coefficients that are generally conservative and close to the stated level. The intervals are also exact in certain situations. For example, interval (6.2.6) is exact when either (i) $\theta_3 = \theta_4 = 0$, or (ii) I and K are fixed and $J \to \infty$, or (iii) J and K are fixed and $I \to \infty$.

Intervals on $\rho_E = \sigma_E^2/\gamma$ are based on (3.4.3) and (3.4.4). Define $\theta_E = [I\theta_1 + J\theta_2 + (IJ-I-J)\theta_3]/\theta_4$ and $\rho_E = IJK/[\theta_E + (K-1)IJ]$. Using (3.4.3) and (3.4.4) with $Q = 4$, $P = 3$, $c_1 = I$, $c_2 = J$, $c_3 = (IJ-I-J)$, and $c_4 = 1$, one can obtain the interval $[L^*; U^*]$ on θ_E. The corresponding interval on ρ_E is $[IJK/(U^* + (K-1)IJ);$ $IJK/(L^* + (K-1)IJ)]$. Birch and Burdick (1989) derived an alternative interval on θ_E that could be used in the same manner to construct an interval on ρ_E.

Example 6.2.2 An objective of the study in Example 6.2.1 was to determine the percentage of variability in blood pH accounted for by differences among the genetic strain of the mice. Interval (6.2.6) can be used to construct an interval on σ_A^2 as a percentage of the total variability in blood pH. Using (6.2.6) with $\alpha = .05$ we compute $L_A = .2931$, $U_A = 5.2381$, and a 90% confidence interval on $\rho_A = \sigma_A^2/(\sigma_A^2 + \sigma_B^2 + \sigma_{AB}^2 + \sigma_E^2)$ is $[.196; .814]$. The point estimate for ρ_A reported by Weir is .42.

6.3 THE BALANCED TWO-FACTOR CROSSED RANDOM MODEL WITHOUT INTERACTION

The data shown in Table 6.3.1 are from Winer (1971, p. 288) and represent ratings made on 6 persons by 4 independent judges in a reliability study. Data of this nature are often used to evaluate the reliability of a population of judges.

The model used to represent these data assumes there is no interaction between the judges and subjects. It can be viewed as either a balanced two-factor crossed random model with no interaction or as a randomized complete block design with random blocks and treatments. The model is represented as

$$Y_{ijk} = \mu + A_i + B_j + E_{ijk}$$
$$i = 1, ..., I; \quad j = 1, ..., J; \quad k = 1, ..., K \geq 1 \quad (6.3.1)$$

where μ is an unknown constant, A_i, B_j, and E_{ijk} are mutually independent normal random variables with means of zero and variances σ_A^2, σ_B^2, and σ_E^2, respectively. This model differs from (6.2.1) in that there is no interaction. In many designs of this type $K = 1$. The analysis of variance for this model is shown in Table 6.3.2 where all the notation is defined in Table 6.2.1. Since S_3^2 and S_4^2 in Table 6.2.1 are both unbiased for σ_E^2 under (6.3.1), they are pooled to form S_P^2 in Table 6.3.2. The computed table for the data in Table 6.3.1 is shown in Table 6.3.3.

6.3.1 Inferences on Variance Components

Interval (3.2.1) with $S_q^2 = S_P^2$ and $n_q = n_P$ provides an exact confidence interval on σ_E^2. Since $\sigma_A^2 = (\theta_1 - \theta_P)/(JK)$ and $\sigma_B^2 = (\theta_2 - \theta_P)/(IK)$, bounds (3.3.1) and (3.3.2) can be used to construct intervals on σ_A^2 and σ_B^2. In particular, (6.2.2) can be used to form a confidence interval on σ_A^2 by replacing S_3^2 and S_P^2 and n_3 with n_P. Simultaneous confidence intervals on σ_A^2 and σ_B^2 when $K = 1$ were developed by Sahai (1974). However, as was illustrated with the nested model in Example 5.2.5, shorter simultaneous intervals can be obtained using the Bonferroni inequality with confidence intervals formed from (3.3.1) and (3.3.2).

Table 6.3.1 Judges Ratings of Persons

Person	Judge 1	Judge 2	Judge 3	Judge 4
1	2	4	3	3
2	5	7	5	6
3	1	3	1	2
4	7	9	9	8
5	2	4	6	1
6	6	8	8	4
Total	23	35	32	24

Table 6.3.2 Analysis of Variance for Balanced Two-Factor Crossed Random Model Without Interaction

SV	DF	SS	MS	EMS
A	$n_1 = I - 1$	$SS1$	S_1^2	$\theta_1 = \sigma_E^2 + JK\sigma_A^2$
B	$n_2 = J - 1$	$SS2$	S_2^2	$\theta_2 = \sigma_E^2 + IK\sigma_B^2$
Pooled Error	$n_P = n_3 + n_4$	$SSP = SS3 + SS4$	S_P^2	$\theta_P = \sigma_E^2$
Total	$IJK - 1$	SST		

Table 6.3.3 Analysis of Variance for Reliability Study

SV	DF	MS
Subjects (A)	5	24.50
Judges (B)	3	5.83
Error	15	1.23
Total	23	

Exact hypothesis tests and MVU estimators for σ_A^2 and σ_B^2 are shown in Table 6.3.4. Tan (1981) has studied the robustness of these tests in the presence of an interaction term when $K = 1$.

Example 6.3.1 Consider the reliability study described in Table 6.3.1 with $I = 6$ subjects and $J = 4$ judges. The covariance between the ratings of two randomly selected judges on a random subject is σ_A^2. Using the ANOVA in Table 6.3.3 a 95% confidence interval on σ_A^2 is [2.06; 36.5]. This interval is computed using (6.2.2) with $S_P^2 = 1.23$ replacing S_3^2 and $n_P = 15$ replacing n_3 in the equation. Also needed for the computations are $G_1 = .6104$, $G_P = .4543$, $H_1 = 5.015$, $H_P = 1.395$, $G_{1P} = -.021$, and $H_{IP} = -.657$.

6.3.2 Inferences on Sums of Variance Components

The total variation of the response variable Y_{ijk} in model (6.3.1) is $\gamma = \sigma_A^2 + \sigma_B^2 + \sigma_E^2$. Confidence intervals on γ are obtained using the results of Section 3.2 with $Q = 3$, $\hat{\gamma} = c_1 S_1^2 + c_2 S_2^2 + c_P S_P^2$, $c_1 = 1/(JK)$, $c_2 = 1/(IK)$, and $c_P = (IJK - I - J)/(IJK)$. An interval on $\sigma_A^2 + \sigma_B^2$ can be obtained from (3.3.3) and (3.3.4) using $P = 2$, $Q = 3$, $c_1 = 1/(JK)$, $c_2 = 1/(IK)$, and $c_P = (I + J)/(IJK)$.

6.3.3 Inferences on Ratios of Variance Components

Since $\lambda_A = \sigma_A^2/\sigma_E^2 = (\theta_1/\theta_P - 1)/(JK)$ and $\lambda_B = \sigma_B^2/\sigma_E^2 = (\theta_2/\theta_P - 1)/(IK)$, interval (3.4.1) can be used to construct exact intervals on λ_A

Table 6.3.4 MVU Estimators and Exact Size α Tests for the Variance Components in Model (6.3.1)

Parameter	MVU Estimator	Hypotheses	Decision Rule: Reject H_o if
σ_A^2	$\dfrac{S_1^2 - S_P^2}{JK}$	$H_o : \sigma_A^2 = 0$ $H_a : \sigma_A^2 > 0$	$\dfrac{S_1^2}{S_P^2} > F_{\alpha:n_1,n_P}$
σ_B^2	$\dfrac{S_2^2 - S_P^2}{IK}$	$H_o : \sigma_B^2 = 0$ $H_a : \sigma_B^2 > 0$	$\dfrac{S_2^2}{S_P^2} > F_{\alpha:n_2,n_P}$

and λ_B. The resulting intervals have the same form as (6.2.4). For example, an exact $1 - 2\alpha$ confidence interval on λ_A is

$$\left[\frac{L^* - 1}{JK} \quad ; \quad \frac{U^* - 1}{JK} \right] \tag{6.3.2}$$

where

$$L^* = \frac{S_1^2}{S_P^2 F_{\alpha:n_1,n_P}} \quad \text{and} \quad U^* = \frac{S_1^2}{S_P^2 F_{1-\alpha:n_1,n_P}}$$

Exact intervals on $\sigma_A^2/(\sigma_A^2 + \sigma_E^2) = \lambda_A/(1 + \lambda_A)$, $\sigma_B^2/(\sigma_B^2 + \sigma_E^2)$ $= \lambda_B/(1 + \lambda_B)$, $\sigma_E^2/(\sigma_A^2 + \sigma_E^2) = 1/(1 + \lambda_A)$, and $\sigma_E^2/(\sigma_B^2 + \sigma_E^2) =$ $1/(1 + \lambda_B)$ are obtained from the intervals on λ_A and λ_B. Broemeling (1969b) derived simultaneous confidence intervals on λ_A and λ_B when $K = 1$ using the bounds in (3.5.1) with S_P^2 and n_P replacing S_Q^2 and n_Q, respectively. Intervals on $\sigma_A^2/\sigma_B^2 = [I(\theta_1 - \theta_P)]/[J(\theta_2 - \theta_P)]$, are formed by replacing n_3 and S_3^2 with n_P and S_P^2 in (6.2.5).

The proportions of variability, $\rho_A = \sigma_A^2/\gamma$ and $\rho_B = \sigma_B^2/\gamma$, where $\gamma = \sigma_A^2 + \sigma_B^2 + \sigma_E^2$, are important ratios in reliability studies. In the context of the data shown in Table 6.3.1, ρ_A is the intraclass correlation of reliability and is often the primary parameter of interest. Although ρ_A and ρ_B cannot be simplified to one of the general forms considered in Chapter 3, approximate intervals for these ratios were developed by Arteaga, Jeyaratnam, and Graybill (1982). Their approximate $1 - 2\alpha$ interval on ρ_A is

$$\left[\frac{IL_A}{IL_A + J} \quad ; \quad \frac{IU_A}{IU_A + J} \right] \tag{6.3.3}$$

where

$$L_A = \frac{S_1^4 - F_{\alpha:n_1,\infty} S_1^2 S_P^2 + (F_{\alpha:n_1,\infty} - F_{\alpha:n_1,n_P})F_{\alpha:n_1,n_P}S_P^4}{(IK - 1)F_{\alpha:n_1,\infty} S_1^2 S_P^2 + F_{\alpha:n_1,n_2} S_1^2 S_2^2} \quad \text{and}$$

$$U_A = \frac{S_1^4 - F_{1-\alpha:n_1,\infty} S_1^2 S_P^2 + (F_{1-\alpha:n_1,\infty} - F_{1-\alpha:n_1,n_P})F_{1-\alpha:n_1,n_P}S_P^4}{(IK - 1)F_{1-\alpha:n_1,\infty} S_1^2 S_P^2 + F_{1-\alpha:n_1,n_2} S_1^2 S_2^2}$$

The recommended $1 - 2\alpha$ confidence interval on ρ_B is

$$\left[\frac{JL_B}{JL_B + 1} \quad ; \quad \frac{JU_B}{JU_B + 1}\right] \tag{6.3.4}$$

where

$$L_B = \frac{S_2^4 - F_{\alpha:n_2,\infty}\,S_2^2 S_P^2 + (F_{\alpha:n_2,\infty} - F_{\alpha:n_2,n_P})F_{\alpha:n_2,n_P}S_P^4}{(JK - 1)F_{\alpha:n_2,\infty}\,S_2^2 S_P^2 + F_{\alpha:n_2,n_1}S_1^2 S_2^2} \quad \text{and}$$

$$U_B = \frac{S_2^4 - F_{1-\alpha:n_2,\infty}\,S_2^2 S_P^2 + (F_{1-\alpha:n_2,\infty} - F_{1-\alpha:n_2,n_P})F_{1-\alpha:n_2,n_P}S_P^4}{(JK - 1)F_{1-\alpha:n_2,\infty}\,S_2^2 S_P^2 + F_{1-\alpha:n_2,n_1}S_1^2 S_2^2}$$

Negative bounds in (6.3.3) and (6.3.4) are replaced with zero. Intervals (6.3.3) and (6.3.4) become exact when either (i) $\theta_P = 0$, or (ii) $J \to \infty$ with I fixed, or (iii) $I \to \infty$ with J fixed. Additionally, (6.3.3) is exact when (iv) $\sigma_A^2 \to \infty$ with σ_B^2 and σ_E^2 fixed and (6.3.4) is exact when (v) $\sigma_B^2 \to \infty$ with σ_A^2 and σ_E^2 fixed. Arteaga, Jeyaratnam, and Graybill used numerical integration to demonstrate that (6.3.3) and (6.3.4) generally maintain the stated levels of confidence.

The ratio $\rho_E = \sigma_E^2/\gamma$ can be expressed in terms of the general ratio considered in Chapter 3. To illustrate, define $\theta_E = (I\theta_1 + J\theta_2)/\theta_P$ and note $\rho_E = (IJK)/(\theta_E + IJK - I - J)$. An approximate $1 - 2\alpha$ confidence interval on θ_E is formed using (3.4.3) and (3.4.4) with $Q = 3, P = 2, c_1 = I, c_2 = J$, and $c_3 = 1$. Lu, Graybill, and Burdick (1987) proposed an alternative interval for θ_E that provides similar results. If $[L_E;U_E]$ is an approximate $1 - 2\alpha$ interval on θ_E, then $[(IJK)/(U_E + IJK - I - J); (IJK)/(L_E + IJK - I - J)]$ is an approximate $1 - 2\alpha$ confidence interval on ρ_E.

Example 6.3.2 Fleiss and Shrout (1978) illustrate how a lower bound on ρ_A can be used to determine the minimum number of independent ratings needed to ensure mean ratings have adequate reliability. In particular, if ρ_A^* is the minimum acceptable value for the reliability coefficient of a mean based on k raters, then k is the smallest integer greater than or equal to $[\rho_A^*(1 - L_{\rho A})]/[L_{\rho A}(1 - \rho_A^*)]$ where $L_{\rho A}$ is a lower bound on ρ_A. Using Table 6.3.3 and the lower bound on ρ_A given in (6.3.3), the computed 95% lower bound on ρ_A is $L_{\rho A} = .329$.

This bound is computed using $F_{.05:5,\infty} = 2.2141$, $F_{.05:5,3} = 9.0135$, $F_{.05:5,15} = 2.9013$. If it is desired to have 95% assurance that the reliability coefficient is at least .70, then $[.70(1 - .329)]/[.329(1 - .70)] = 4.8$ and at least five ratings are needed in any future study.

6.4 THE BALANCED THREE-FACTOR CROSSED RANDOM MODEL WITH INTERACTION

As illustrated is Sections 6.2 and 6.3, confidence intervals for many of the parameters of interest in crossed designs are obtained from the general results of Chapter 3. To further illustrate this point we now consider the three-factor crossed random model with interaction. This model is written as

$$Y_{ijkm} = \mu + A_i + B_j + C_k + (AB)_{ij} +$$
$$(AC)_{ik} + (BC)_{jk} + (ABC)_{ijk} + E_{ijkm}$$
$$i = 1, \ldots, I; \quad j = 1, \ldots, J; \qquad (6.4.1)$$
$$k = 1, \ldots, K; \quad m = 1, \ldots, M$$

where μ is a constant, A_i, B_j, C_k, $(AB)_{ij}$, $(AC)_{ik}$, $(BC)_{jk}$, $(ABC)_{ijk}$, and E_{ijkm} are jointly independent normal random variables with means of zero and variances σ_A^2, σ_B^2, σ_C^2, σ_{AB}^2, σ_{AC}^2, σ_{BC}^2, σ_{ABC}^2, and σ_E^2, respectively. The ANOVA for model (6.4.1) is shown in Table 6.4.1.

6.4.1 Inferences on Variance Components

Under the assumptions of model (6.4.1), $n_q S_q^2/\theta_q$ are jointly independent chi-squared random variables with n_q degrees of freedom, respectively for $q = 1, \ldots, 8$. These distributional assumptions allow the use of the results of Section 3.3 to construct confidence intervals on the individual variance components. In particular, (3.3.3) and (3.3.4) are used to form intervals on $\sigma_A^2 = (\theta_1 + \theta_7 - \theta_4 - \theta_5)/(JKM)$, $\sigma_B^2 = (\theta_2 + \theta_7 - \theta_4 - \theta_6)/(IKM)$, and $\sigma_C^2 = (\theta_3 + \theta_7 - \theta_5 - \theta_6)/(IJM)$. To illustrate, the lower bound in a $1 - \alpha$ upper confidence interval on σ_A^2 based on (3.3.3) is

$$\frac{S_1^2 + S_7^2 - S_4^2 - S_5^2 - \sqrt{V_L}}{JKM} \qquad (6.4.2)$$

Table 6.4.1 Analysis of Variance for Balanced Random Three-Factor Crossed Design

SV	DF	MS	EMS
A	$n_1 = I - 1$	S_1^2	$\theta_1 = \sigma_E^2 + M\sigma_{ABC}^2 + JM\sigma_{AC}^2 + KM\sigma_{AB}^2 + JKM\sigma_A^2$
B	$n_2 = J - 1$	S_2^2	$\theta_2 = \sigma_E^2 + M\sigma_{ABC}^2 + KM\sigma_{AB}^2 + IM\sigma_{BC}^2 + IKM\sigma_B^2$
C	$n_3 = K - 1$	S_3^2	$\theta_3 = \sigma_E^2 + M\sigma_{ABC}^2 + IM\sigma_{BC}^2 + JM\sigma_{AC}^2 + IJM\sigma_C^2$
AB	$n_4 = (I - 1)(J - 1)$	S_4^2	$\theta_4 = \sigma_E^2 + M\sigma_{ABC}^2 + KM\sigma_{AB}^2$
AC	$n_5 = (I - 1)(K - 1)$	S_5^2	$\theta_5 = \sigma_E^2 + M\sigma_{ABC}^2 + JM\sigma_{AC}^2$
BC	$n_6 = (J - 1)(K - 1)$	S_6^2	$\theta_6 = \sigma_E^2 + M\sigma_{ABC}^2 + IM\sigma_{BC}^2$
ABC	$n_7 = (I - 1)(J - 1)(K - 1)$	S_7^2	$\theta_7 = \sigma_E^2 + M\sigma_{ABC}^2$
ERROR	$n_8 = (M - 1)IJK$	S_8^2	$\theta_8 = \sigma_E^2$

where

$$V_L = G_1^2 S_1^4 + G_7^2 S_7^4 + H_4^2 S_4^4 + H_5^2 S_5^4$$
$$+ G_{14} S_1^2 S_4^2 + G_{15} S_1^2 S_5^2 + G_{74} S_7^2 S_4^2 + G_{75} S_7^2 S_5^2 + G_{17}^* S_1^2 S_7^2$$

$$G_q = 1 - \frac{1}{F_{\alpha:n_q,\infty}} \qquad (q = 1, 7)$$

$$H_r = \frac{1}{F_{1-\alpha:n_r,\infty}} - 1 \qquad (r = 4, 5)$$

$$G_{qr} = \frac{(F_{\alpha:n_q,n_r} - 1)^2 - G_q^2 F_{\alpha:n_q,n_r}^2 - H_r^2}{F_{\alpha:n_q,n_r}} \qquad (q = 1, 7; r = 4, 5) \qquad \text{and}$$

$$G_{17}^* = \left(1 - \frac{1}{F_{\alpha:n_1+n_7,\infty}}\right)^2 \frac{(n_1 + n_7)^2}{n_1 n_7} - \frac{G_1^2 n_1}{n_7} - \frac{G_7^2 n_7}{n_1}$$

The bounds in (3.3.1) and (3.3.2) are used to form intervals on $\sigma_{AB}^2 = (\theta_4 - \theta_7)/(KM)$, $\sigma_{AC}^2 = (\theta_5 - \theta_7)/(JM)$, $\sigma_{BC}^2 = (\theta_6 - \theta_7)/(IM)$, and $\sigma_{ABC}^2 = (\theta_7 - \theta_8)/M$. Interval (3.2.1) provides an exact interval on σ_E^2.

Exact hypothesis tests for $H_o : \sigma_{AB}^2 = 0$, $H_o : \sigma_{AC}^2 = 0$, $H_o : \sigma_{BC}^2 = 0$, and $H_o : \sigma_{ABC}^2 = 0$ are based on the ratios S_4^2/S_7^2, S_5^2/S_7^2, S_6^2/S_7^2, and S_7^2/S_8^2, respectively. There are no exact tests for the hypotheses $H_o : \sigma_A^2 = 0$, $H_o : \sigma_B^2 = 0$, and $H_o : \sigma_C^2 = 0$ that are based on the statistics in the ANOVA table. Consider, for example, the test of $H_o : \sigma_A^2 = 0$. There is no ratio of the S_q^2 that has an exact F-distribution under the null hypothesis $\sigma_A^2 = 0$. In this situation, many investigators perform an approximate F-test using a ratio of linear combinations of the S_q^2. For example, Scheffé (1959, p. 248) recommends the ratio $F_1 = S_1^2/(S_4^2 + S_5^2 - S_7^2)$ and Cochran (1951) the ratio $F_2 = (S_1^2 + S_7^2)/(S_4^2 + S_5^2)$ for testing $H_o : \sigma_A^2 = 0$. If $H_o : \sigma_A^2 = 0$ is true, the expected values of the numerators and denominators are equal in both F_1 and F_2. Each ratio is then treated like an F-statistic with degrees of freedom estimated using the Satterthwaite approximation. Davenport and Webster (1973) compared F_1 and F_2 for testing $H_o : \sigma_A^2 = 0$ and found their performance to be similar with a slight preference for F_2 when degrees of freedom are small. Myers and Howe (1971) considered an alternative to the Satterthwaite approximation for estimating the degrees of freedom of F_1 and F_2, but Davenport (1975) showed this alternative provides more liberal tests. An alternative test of $H_o : \sigma_A^2 = 0$ was proposed by Jeyaratnam and Graybill (1980). They considered a test based on a lower bound of $\sigma_A^2 = (\theta_1 + \theta_7 - \theta_4 - \theta_5)/(JKM)$. Since $\sigma_A^2 = 0$ when H_0 is true, the null hypothesis is rejected if the lower bound on σ_A^2 is positive. The lower bound proposed by Jeyaratnam and Graybill is different than the one shown in (6.4.2). Thus, another alternative for testing $H_o : \sigma_A^2 = 0$ is to compute (6.4.2) and reject H_o if it is positive. We note that this test based on (6.4.2) is equivalent to a test based on the lower bound of $(\theta_1 + \theta_7)/(\theta_4 + \theta_5)$ using (3.4.3). In this later test, H_o is rejected if the lower bound exceeds one.

Birch, Burdick, and Ting (1990) conducted a simulation study to compare several tests for $H_o : \sigma_A^2 = 0$. In particular, they studied the tests based on F_2 using a modified Satterthwaite approximation, the Jeyaratnam-Graybill lower bound on σ_A^2, and the lower bound in (6.4.2). Their results suggest that both the Satterthwaite and Jeyaratnam-Graybill tests can have levels of significance greater than the stated level. The test based on (6.4.2), on the other hand, is generally conservative, but does not suffer an undue loss of power. To summarize, we recommend the lower bound on σ_A^2 in (6.4.2) for testing the hypothesis $H_o : \sigma_A^2 = 0$. If the lower bound is positive, the null hypothesis is rejected. The lower bounds on σ_B^2 and σ_C^2 formed from (3.3.3) can likewise be used for testing $H_o : \sigma_B^2 = 0$ and $H_o : \sigma_C^2 = 0$.

We note that if one is willing to make assumptions concerning the variance components associated with the interactions, exact tests can be obtained for the main effects. For example, if it is assumed that $\sigma_{AB}^2 = 0$, then S_1^2/S_5^2 provides an exact test statistic for $H_o : \sigma_A^2 = 0$. However, Naik (1974) points out that a test conducted in this manner will have a greater test size than the nominal level. Naik also provides tests with test sizes no greater than the stated level for the model where $\sigma_{ABC}^2 = 0$. Seifert (1981) provides exact tests for the main effects in model (6.4.1) that are of the Bartlett-Scheffé type. However, these tests involve arbitrary linear combinations of cell means. If different linear combinations are selected, the value of the test statistics will change, and the test statistic is not unique for a given set of data. Additionally, these exact tests will not be as powerful as the approximate tests recommended in this section.

Example 6.4.1   Johnson and Leone (1977, p. 861) report data from an experiment concerning the melting point of a particular chemical.

The experiment was performed with three analysts using three thermometers in three separate weeks. All effects were crossed and random. Table 6.4.2 reports the results of the experiment. Although $M = 1$ and no degrees of freedom exist for estimating σ_E^2, confidence intervals can be computed on σ_A^2, σ_B^2, and σ_C^2 using the methods discussed in this section. To illustrate, a lower bound on σ_A^2, the temperature variance component, is computed using (6.4.2). The 90% lower bound based on Table 6.4.2 is .062 where $J = 3$, $K = 3$, $M = 1$, $G_1 = .566$, $G_7 = .401$, $H_4 = H_5 = 2.76$, $G_{14} = G_{15} = -.591$, $G_{74} = G_{75} = -.356$, $G_{17}^* = .153$, and $V_L = 4.903$. Since the lower bound is positive, we reject $H_o : \sigma_A^2 = 0$ with a .10 level of significance.

6.4.2 Inferences on Sums of Variance Components

A confidence interval on the total variation $\gamma = \sigma_A^2 + \sigma_B^2 + \sigma_C^2 + \sigma_{AB}^2 + \sigma_{AC}^2 + \sigma_{BC}^2 + \sigma_{ABC}^2 + \sigma_E^2$ is formed using the results of Section 3.2. To determine the appropriate values of the c_q, simply sum the ANOVA estimators of the individual variance components. In this case,

$$Q = 8; \quad c_1 = I; \quad c_2 = J; \quad c_3 = K;$$
$$c_4 = IJ - I - J; \quad c_5 = IK - I - K;$$
$$c_6 = JK - K - J;$$
$$c_7 = (I - 1)(J - 1)(K - 1) + 1; \quad \text{and}$$
$$c_8 = IJK(M - 1)$$

provide an interval on $(IJKM)\,\gamma$.

Table 6.4.2 ANOVA for Thermometer Data

SV	DF	MS
Thermometer (A)	2	3.444
Analyst (B)	2	1.083
Week (C)	2	.3333
Thermometer × analyst	4	.5278
Thermometer × week	4	.2778
Analyst × week	4	.4167
Thermometer × analyst × week	8	.1319
Error	0	0
Total	26	

6.4.3   Inferences on Ratios of Variance Components

The set of five expected means squares, $\theta_4, ..., \theta_8$, have the same form as the set of expected mean squares in the two-factor crossed random model with interaction. As such, ratios that involve only σ_{AB}^2, σ_{AC}^2, σ_{BC}^2, σ_{ABC}^2, and σ_E^2 can be constructed using the results of Section 6.2.3. For example, $\sigma_{AB}^2/\sigma_{AC}^2 = J(\theta_4 - \theta_7)/[K(\theta_5 - \theta_7)]$ and (6.2.5) with appropriate substitutions can be used to form an interval on this ratio. Ratios that involve σ_A^2, σ_B^2, and σ_C^2 are more complex and no general results are presently available.

6.5   UNBALANCED TWO-FACTOR CROSSED RANDOM MODELS WITH INTERACTION

We now consider crossed designs in which there are an unequal number of observations in the cells. The first design considered is (6.2.1) when sample sizes are unequal but no cells are missing. The data shown in Table 6.5.1 correspond to this design. They are a subset of data reported by Milliken and Johnson (1984, p. 265) for a manufacturing plant where it was of interest to study the efficiency of workers in assembly lines. In the experiment three assembly sites in the plant were randomly selected and four workers were randomly selected from the pool of factory workers. During the course of the experiment, workers often worked at particular sites more than once so that most cells had more than one observation. The response variable represents an efficiency score.

The unbalanced two-factor crossed random model with interaction and no missing cells is written as

$$Y_{ijk} = \mu + A_i + B_j + (AB)_{ij} + E_{ijk} \qquad (6.5.1)$$
$$i = 1, ..., I; \qquad j = 1, ..., J; \qquad k = 1, ..., K_{ij} > 0$$

where μ, A_i, B_j, $(AB)_{ij}$, and E_{ijk} are defined in (6.2.1). For example, the data shown in Table 6.5.1 has $I = 3$ sites, $J = 4$ workers, $K_{11} = 3$, and $K_{23} = 1$. Unlike the balanced design, there is no unique partitioning of the total sums of squares that provides a set of independently distributed chi-squared random variables. However, as in previous chapters, the unweighted sums of squares associated with the effects in (6.5.1) provide a set of mean squares that work well in the

Table 6.5.1   Assembly Line Data

Site	Worker			
	1	2	3	4
1	100.6	110.0	100.0	98.2
	106.8	105.8	102.5	99.5
	100.6		97.6	
			98.7	
			98.7	
2	92.3	103.2	96.4	108.0
	92.0	100.5		108.9
	97.2	100.2		107.9
	93.9	97.7		
	93.0			
3	96.9	92.5	86.8	94.4
	96.1	85.9		93.0
	100.8	85.2		91.0
		89.4		
		88.7		

Chapter 3 equations. Table 6.5.2 reports these unweighted sums of squares in the form of an ANOVA table.

The means in Table 6.5.2 are defined as $\bar{Y}_{ij*} = \Sigma_k Y_{ijk}/K_{ij}$, $\bar{Y}_{i*}^* = \Sigma_j \bar{Y}_{ij*}/J$, $\bar{Y}_{*j}^* = \Sigma_i \bar{Y}_{ij*}/I$, $\bar{Y}_*^* = \Sigma_i \bar{Y}_{i*}^*/I = \Sigma_j \bar{Y}_{*j}^*/J = \Sigma_i \Sigma_j \bar{Y}_{ij*}/(IJ)$, and $\bar{K}_H = (IJ)/\Sigma_i \Sigma_j 1/K_{ij}$. The mean square S_4^2 is independent of S_{1U}^2, S_{2U}^2, and S_{3U}^2. The random variable $n_4 S_4^2/\theta_4$ has a chi-squared distribution with n_4 degrees of freedom. In general S_{1U}^2, S_{2U}^2, and S_{3U}^2 are not jointly independent, and none of these mean squares have scaled chi-squared distributions. If $K_{ij} = K$, the design is balanced and $\bar{K}_H = K$, $S_{1U}^2 = S_1^2$, $S_{2U}^2 = S_2^2$, and $S_{3U}^2 = S_3^2$ where S_1^2, S_2^2 and S_3^2 are defined in Table 6.2.1.

The only functions for which exact intervals can be obtained are σ_E^2 and σ_{AB}^2/σ_E^2. Since $n_4 S_4^2/\theta_4$ is a chi-squared random variable, interval (3.2.1) provides an exact $1 - 2\alpha$ confidence interval on σ_E^2. The Wald interval described in Appendix B provides an exact interval on σ_{AB}^2/σ_E^2. The required computations are illustrated in Example 6.5.1. The Wald interval simplifies to (6.2.4) when $K_{ij} = K$.

Table 6.5.2 Analysis of Variance with Unweighted SS for Unbalanced Two-Factor Crossed Random Model with Interaction

SV	DF	SS	MS	EMS
A	$n_1 = I - 1$	$SS1U = J\bar{K}_H \sum_i (\bar{Y}^*_{i*} - \bar{Y}^*_{*})^2$	S^2_{1U}	$\theta_{1U} = \sigma^2_E + \bar{K}_H \sigma^2_{AB} + J\bar{K}_H \sigma^2_A$
B	$n_2 = J - 1$	$SS2U = I\bar{K}_H \sum_j (\bar{Y}^*_{*j} - \bar{Y}^*_{*})^2$	S^2_{2U}	$\theta_{2U} = \sigma^2_E + \bar{K}_H \sigma^2_{AB} + I\bar{K}_H \sigma^2_B$
AB	$n_3 = (I-1)(J-1)$	$SS3U = \bar{K}_H \sum_i \sum_j (\bar{Y}_{ij*} - \bar{Y}^*_{i*} - \bar{Y}^*_{*j} + \bar{Y}^*_{*})^2$	S^2_{3U}	$\theta_{3U} = \sigma^2_E + \bar{K}_H \sigma^2_{AB}$
Error	$n_4 = \sum_i \sum_j K_{ij} - IJ$	$SS4 = \sum_i \sum_j \sum_k (Y_{ijk} - \bar{Y}_{ij*})^2$	S^2_4	$\theta_4 = \sigma^2_E$

The Wald interval on σ_{AB}^2/σ_E^2 provides an exact test of $H_o : \sigma_{AB}^2 = 0$. The same test was derived by Spjøtvoll (1968) and Thomsen (1975). Seely and El-Bassiouni (1983, p. 200) show this equivalence. Using the notation of Appendix B, this test is equivalent to rejecting $H_o : \sigma_{AB}^2 = 0$ if $G(0) = f\Sigma_i^r t_i^2/(rSSE) < F_{\alpha:r,f}$. In the notation of this chapter, $SSE/f = S_4^2$, $f = n_4$, $r = n_3$, and $\Sigma_i^r t_i^2$ corresponds to the interaction sum of squares in the fixed effects model discussed in Graybill (1976, p. 577) and Searle (1987, p. 104, equation (76)).

The results of Section 6.2 can be used for constructing approximate intervals on other functions of the variance components. Comparison of the expected mean squares in Tables 6.2.1 and Table 6.5.2 shows the only difference is that $\tilde{K}_H$ replaces K in Table 6.5.2. Thus, the approximate confidence intervals presented in Section 6.2 can be used in an unbalanced design with no missing cells by replacing S_1^2, S_2^2, S_3^2, and K, with S_{1U}^2, S_{2U}^2, S_{3U}^2, and $\tilde{K}_H$, respectively. Although this approach violates the chi-squaredness and independence assumptions made in Section 6.2, these approximate intervals seem to perform well. Hernandez (1991) provides simulations for the intervals in Sections 6.2.1 and 6.2.2 that suggest the lack of independence and chi-squaredness have canceling effects on the confidence coefficient. Thus, the intervals perform as well in unbalanced designs as they do in balanced designs.

Several methods have been proposed for testing $H_o : \sigma_A^2 = 0$ and $H_o: \sigma_B^2 = 0$. Exact tests for these hypotheses were proposed by Spjøtvoll (1968) and Thomsen (1975) under the assumption that $\sigma_{AB}^2 = 0$. The test proposed by Thomsen with no missing cells is the Wald test. This test is different than the one proposed by Spjøtvoll. However, as demonstrated by Seely and El-Bassiouni (1983), it is not possible to construct a Wald test for either $H_o : \sigma_A^2 = 0$ or $H_o : \sigma_B^2 = 0$ without assuming $\sigma_{AB}^2 = 0$. Khuri and Littell (1987) provide exact tests of $H_o : \sigma_A^2 = 0$ and $H_o : \sigma_B^2 = 0$ that do not require $\sigma_{AB}^2 = 0$. However, these tests require a non-unique partitioning of the error sums of squares. Since no unique exact test has been proposed for testing either $H_o : \sigma_A^2 = 0$ or $H_o : \sigma_B^2 = 0$ without assuming $\sigma_{AB}^2 = 0$, we recommend approximate tests based on lower bounds formed using the unweighted sums of squares. This is equivalent to rejecting $H_o : \sigma_A^2 = 0$ if $S_{1U}^2/S_{3U}^2 > F_{\alpha:n_1,n_3}$ and rejecting $H_o : \sigma_B^2 = 0$ if $S_{2U}^2/S_{3U}^2 > F_{\alpha:n_2,n_3}$.

Hernandez (1991) shows these tests generally maintain the test size and have comparable power to the exact test of Khuri and Littell (1987).

Tan, Tabatabai, and Balakrishnan (1988) report tests for $H_o : \sigma_A^2 = 0$, $H_o : \sigma_B^2 = 0$, and $H_o : \sigma_{AB}^2 = 0$ that can be used when σ_E^2 is not equal in all cells. These tests are also based on the unweighted sums of squares. Hussein and Milliken (1978a) provide tests for $H_o : \sigma_A^2 = 0$ and $H_o : \sigma_B^2 = 0$ in a heteroscedastic situation where $\sigma_{AB}^2 = 0$.

Example 6.5.1   We illustrate the discussion in this section by considering the data in Table 6.5.1. A lower bound on σ_A^2 is obtained by noting $\sigma_A^2 = (\theta_{1U} - \theta_{3U})/(J\bar{K}_H)$. Thus, (6.2.2) can be used to compute a bound on σ_A^2 by replacing K with $\bar{K}_H$, S_1^2 with S_{1U}^2, and S_3^2 with S_{3U}^2. In order to compute the unweighted mean squares, S_{1U}^2 and S_{3U}^2, we perform an analysis of variance on the cell means. For example, the following SAS code computes the cell means, and then uses PROC GLM to compute the unweighted mean squares.

```
DATA EX651;
INPUT SITE WORKER Y;
CARDS;
1 1 100.6
1 1 106.8
. . . .
DATA GOES HERE
. . . .
3 4 91.0
PROC SORT;
BY SITE WORKER;
*CALCULATE THE CELL MEANS AND PLACE THEM IN A NEW DATASET CALLED USS;
PROC MEANS NOPRINT;
BY SITE WORKER;
VAR Y;
OUTPUT OUT=USS MEAN=YIJBAR;
*RUN THE ANOVA ON THE CELL MEANS.  THE SUM OF SQUARES FOR SITE
IS SS1U/KBARH.   THE SUM OF SQUARES FOR INTERACTION IS SS3U/KBARH;
PROC GLM DATA=USS;
CLASSES SITE WORKER;
MODEL YIJBAR=SITE WORKER SITE*WORKER;
```

The sum of squares reported on the SAS printout for Site corresponds to $n_1 S_{1U}^2 / \bar{K}_H = 253.07$. Similarly, the sum of squares reported for the Site × Worker interaction is $n_3 S_{3U}^2 / \bar{K}_H = 191.30$. By using (6.2.2) with the appropriate substitutions, the approximate 90% lower bound on σ_A^2 is 2.33. In the computations, $\bar{K}_H = 2.315$, $G_1 = .5657$, $H_3 = 1.722$, and $G_{13} = -.2127$. Since the lower bound is positive, we reject $H_o : \sigma_A^2 = 0$ with an approximate test size of $\alpha = .10$.

As described in Appendix B, the exact Wald interval on $\sigma_{AB}^2 / \sigma_E^2$ is computed by treating the random effects A_i and B_j as if they are fixed effects. This results in the design matrix for fixed effects

$$
X = \begin{bmatrix}
1 & 1 & 0 & 0 & 1 & 0 & 0 & 0 \\
1 & 1 & 0 & 0 & 0 & 1 & 0 & 0 \\
\vdots & \vdots & \vdots & \vdots & \vdots & \vdots & \vdots & \vdots \\
1 & 0 & 0 & 1 & 0 & 0 & 0 & 1
\end{bmatrix}
\begin{matrix}
\}K_{11} = 3 \text{ rows} \\
\}K_{12} = 2 \text{ rows} \\
\vdots \\
\}K_{34} = 3 \text{ rows}
\end{matrix} \quad (6.5.2)
$$

The matrix X has 37 rows and 8 columns. The design matrix of the random effects is formed by multiplying the columns of X corresponding to A_i and B_j. This provides the design matrix

$$
Z = \begin{bmatrix}
1 & 0 & 0 & 0 & 0 & 0 & 0 & 0 & 0 & 0 & 0 & 0 \\
0 & 1 & 0 & 0 & 0 & 0 & 0 & 0 & 0 & 0 & 0 & 0 \\
\vdots & \vdots & \vdots & \vdots & \vdots & \vdots & \vdots & \vdots & \vdots & \vdots & \vdots & \vdots \\
0 & 0 & 0 & 0 & 0 & 0 & 0 & 0 & 0 & 0 & 0 & 1
\end{bmatrix}
\begin{matrix}
\}K_{11} = 3 \text{ rows} \\
\}K_{12} = 2 \text{ rows} \\
\vdots \\
\}K_{34} = 3 \text{ rows}
\end{matrix} \quad (6.5.3)
$$

The matrix Z has 37 rows and 12 columns. For the data contained in Table 6.5.1, $r = 6$, $p^* = 6$, $f = 25$, $\Delta_1 = 1.2196$, and $\Delta_6 = 4.1672$. After 13 iterations of the bisection method algorithm, these data yield the exact 90% confidence interval [2.02; 22.2]. Since the lower bound of this interval is positive, we reject $H_o : \sigma_{AB}^2 = 0$. This is an exact test with size $\alpha = .05$.

As discussed in Appendix B, a conservative interval on $\sigma_{AB}^2 / \sigma_E^2$ is provided by (B.7). For the design considered in this section it can be shown that $\text{Min}(K_{ij}) \leq \Delta_1 \leq \cdots \leq \Delta_r \leq \text{Max}(K_{ij})$ and $\Sigma_i t_i^2 / \Delta_i = n_3 S_{3U}^2 / \bar{K}_H$. Thus, replacement of Δ_1 and Δ_r with $\text{Min}(K_{ij})$ and Max

(K_{ij}) in L_1 and U_r, respectively, provides the $1 - 2\alpha$ conservative interval

$$[L_m;\ U_M] \tag{6.5.4}$$

where

$$L_m = \frac{S_{3U}^2}{\bar{K}_H S_4^2 F_{\alpha:n_3,n_4}} - \frac{1}{m};\quad U_M = \frac{S_{3U}^2}{\bar{K}_H S_4^2 F_{1-\alpha:n_3,n_4}} - \frac{1}{M}$$

$$m = \text{Min}(K_{ij})\quad \text{and}\quad M = \text{Max}(K_{ij})$$

Interval (6.5.4) reduces to (6.2.4) when $K_{ij} = K$. For the data in Table 6.5.1, this conservative 90% interval is [1.36; 22.3]. This interval is only slightly wider than the exact interval.

As a final note, the unweighted sums of squares cannot be defined when some $K_{ij} = 0$. Kazempour and Graybill (1991a) considered constructing intervals on ρ_A and ρ_B in models with missing cells, and recommended an alternative to the unweighted sums of squares. In particular, their proposed sums of squares are equivalent to the Type II sums of squares in PROC GLM of SAS©. Intervals on ρ_A and ρ_B were formed using (6.3.3) and (6.3.4) with appropriate substitutions. Section 6.6.2 provides additional discussion of this approach in the without interaction model.

6.6 UNBALANCED TWO-FACTOR CROSSED RANDOM MODELS WITHOUT INTERACTION

6.6.1 Designs with No Missing Cells

Consider the data in Table 6.5.1 and assume that no interaction exists between worker and site. The unbalanced two-factor crossed random model with no interaction and no missing cells is written as

$$Y_{ijk} = \mu + A_i + B_j + E_{ijk} \tag{6.6.1}$$
$$i = 1, ..., I;\quad j = 1, ..., J;\quad k = 1, ..., K_{ij} > 0$$

where μ, A_i, B_j, and E_{ijk} are defined in (6.3.1). By setting $\sigma_{AB}^2 = 0$ in Table 6.5.2, it is seen that under model (6.6.1), $E(S_{1U}^2) = \theta_{1U} = \sigma_E^2$

$+ J\bar{K}_H\sigma_A^2$, $E(S_{2U}^2) = \theta_{2U} = \sigma_E^2 + I\bar{K}_H\sigma_B^2$, $E(S_{3U}^2) = \theta_{3U} = \sigma_E^2$, and $E(S_4^2) = \theta_4 = \sigma_E^2$. Since S_{3U}^2 and S_4^2 now have the same expected values, we follow the approach of Section 6.3 and pool them to form $S_{PU}^2 = (n_3 S_{3U}^2 + n_4 S_4^2)/n_P$ where $E(S_{PU}^2) = \theta_{PU} = \sigma_E^2$ and $n_P = n_3 + n_4$. However, S_{3U}^2 is not independent of S_{1U}^2 and S_{2U}^2 even when $\sigma_{AB}^2 = 0$. Thus, S_{PU}^2 will not in general be independent of S_{1U}^2 and S_{2U}^2. Additionally, neither $n_1 S_{1U}^2/\theta_{1U}$, $n_2 S_{2U}^2/\theta_{2U}$, nor $n_P S_{PU}^2/\theta_{PU}$ in general have chi-squared distributions.

Following the approach of Section 6.5, we recommend the unweighted sums of squares substitutions in the balanced model equations of Section 6.3 for constructing confidence intervals on functions of the variance components in model (6.6.1). In particular, S_{1U}^2, S_{2U}^2, S_{PU}^2, and $\bar{K}_H$ replace S_1^2, S_2^2, S_P^2, and K, respectively, in these formulas. Hernandez (1991) and Srinivasan (1986) provide simulations that indicate these substitutions provide good intervals for the functions considered in Sections 6.3.1 and 6.3.2. Srinivasan and Graybill (1991) made these substitutions in (6.3.3) and (6.3.4) to construct good approximate intervals on $\rho_A = \sigma_A^2/\gamma$ and $\rho_B = \sigma_B^2/\gamma$ where $\gamma = \sigma_A^2 + \sigma_B^2 + \sigma_E^2$. They also considered modifications to the degrees of freedom, but found that that the modifications are not generally needed.

Exact intervals on σ_A^2/σ_E^2 and σ_B^2/σ_E^2 can be obtained from the Wald interval described in Appendix B. These intervals provide exact tests for $H_o : \sigma_A^2 = 0$ and $H_o : \sigma_B^2 = 0$. These exact tests were also proposed by Thomsen (1975, p. 260, equation 2.7) and correspond to the F-tests for main effects in the fixed effects two-factor model with no interaction discussed in Graybill (1976, Section 14.4) and Searle (1987, p. 120, equations (106) and (107)). Spjøtvoll (1968, p. 41) proposed an alternative exact test. Mathew and Sinha (1988b) report optimal tests for $H_o : \sigma_A^2 = 0$ and $H_o : \sigma_B^2 = 0$ when a certain structure of unbalancedness exists.

Example 6.6.1 To illustrate the computations described in this section we consider the data in Table 6.5.1. We assume the no interaction model (6.6.1) and construct a lower bound on σ_A^2. To form this bound note that $\sigma_A^2 = (\theta_{1U} - \theta_{PU})/(J\bar{K}_H)$. Thus, the bound is computed using (6.2.2) and replacing S_1^2 with S_{1U}^2, S_3^2 with S_{PU}^2, and n_3 with n_P. As demonstrated in Example 6.5.1, the unweighted mean squares are computed by performing an analysis of variance on the cell means. The computed 90% lower bound on σ_A^2 is 11.7. For these calculations $S_{1U}^2 = 292.9$, $S_{PU}^2 = 18.66$, and $\bar{K}_H = 2.315$.

The exact Wald interval is now computed for σ_A^2/σ_E^2. To compute this interval, the B_j effects are treated as fixed and combined with μ to define the matrix of fixed effects. This matrix is

$$
X = \begin{bmatrix}
1 & 1 & 0 & 0 & 0 \\
1 & 0 & 1 & 0 & 0 \\
1 & 0 & 0 & 1 & 0 \\
1 & 0 & 0 & 0 & 1
\end{bmatrix}
\begin{matrix}
\}K_{*1} = 11 \text{ rows} \\
\}K_{*2} = 11 \text{ rows} \\
\}K_{*3} = 7 \text{ rows} \\
\}K_{*4} = 8 \text{ rows}
\end{matrix}
\qquad (6.6.2)
$$

The matrix X has 37 rows and 5 columns. The design matrix of the random effects corresponds to the A_i effects. This matrix is

$$
Z = \begin{bmatrix}
1 & 0 & 0 \\
0 & 1 & 0 \\
0 & 0 & 1
\end{bmatrix}
\begin{matrix}
\}K_{1*} = 12 \text{ rows} \\
\}K_{2*} = 13 \text{ rows} \\
\}K_{3*} = 12 \text{ rows}
\end{matrix}
\qquad (6.6.3)
$$

The matrix Z has 37 rows and 3 columns. For the data contained in Table 6.5.1, $r = 2$, $p^* = 4$, $f = 31$, $\Delta_1 = 10.075$, and $\Delta_2 = 12.318$. After 8 iterations of the bisection method algorithm, these data yield the exact 90% confidence interval [.33; 27.1]. Since the lower bound of this interval is positive, we reject $H_o : \sigma_A^2 = 0$. This is an exact test with size $\alpha = .05$.

6.6.2 Designs with Missing Cells

The unweighted sums of squares cannot be defined when some $K_{ij} = 0$. In situations where the design is connected (see, e.g., Graybill, 1976, p. 549), one can replace the unweighted sums of squares with the "adjusted" sums of squares used for testing main effects in the fixed effects model described in Graybill (1976, Section 14.4) and Searle (1987, pages 153–154, equations (49)–(51)). These sums of squares correspond to the Type II, Type III, or Type IV sums of squares reported in PROC GLM of SAS©. We denote the associated mean squares as S_{1W}^2, S_{2W}^2, and S_{PW}^2, corresponding to the A main effect, the B main effect, and the pooled error, respectively. For model (6.6.1) $E(S_{1W}^2) = \theta_{1W} = \sigma_E^2 + J\bar{K}_{1W}\sigma_A^2$, $E(S_{2W}^2) = \theta_{2W} = \sigma_E^2 + I\bar{K}_{2W}\sigma_B^2$, and $E(S_{PW}^2) = \sigma_E^2$ where $\bar{K}_{1W} = [K_{**} - \Sigma_i\Sigma_j K_{*j}^{-1}K_{ij}^2]/[J(I - 1)]$, and $\bar{K}_{2W} = [K_{**} - \Sigma_i\Sigma_j K_{i*}^{-1}K_{ij}^2]/[I(J - 1)]$. The statistics S_{1W}^2/S_{PW}^2 and S_{2W}^2/S_{PW}^2 provide the Wald tests for $H_o : \sigma_A^2 = 0$ and $H_o : \sigma_B^2 = 0$, respectively. To illustrate, (6.2.2) can be used for constructing a confidence interval on σ_A^2 by replacing S_1^2, S_3^2, and K with S_{1W}^2, S_{PW}^2, and $\bar{K}_{1W}$, respectively. Kazempour and Graybill (1992) used this approach for constructing intervals on σ_A^2 and σ_B^2, although they used the interval proposed by Howe (1974) instead of (6.2.2). They also used a Satterthwaite approximation to adjust the degrees of freedom associated with S_{1W}^2 and S_{2W}^2. Kazempour and Graybill (1989, 1991b) used this same approach for constructing confidence intervals on ρ_A, ρ_B, and ρ_E. The interval on ρ_E was based on Lu et al. (1987).

Example 6.6.2 Consider the data in Table 6.5.1 but assume the values 96.4 in cell (2,3) and 86.8 in cell (3,3) are missing. We assume the no interaction model in (6.6.1) and use (6.2.2) with appropriate substitutions to construct a 90% lower bound on σ_A^2. The computations yield $S_{1W}^2 = 288.45$, $S_{PW}^2 = 25.110$, $n_1 = 2$, $n_P = 29$, and $\bar{K}_{1W} = 2.40625$. This provides the approximate 90% confidence interval on σ_A^2, [7.27; 581.5]. The 90% Wald confidence interval on σ_A^2/σ_E^2 is [.266; 24.30]. This interval is computed as described in Example 6.6.1.

6.7 BALANCED INCOMPLETE BLOCK RANDOM MODELS

The balanced incomplete block is a special case of the model considered in Section 6.6.2. In particular, consider a balanced incomplete block design where I treatments are replicated in r of J blocks with m plots per block. Each pair of treatments occurs in exactly $\lambda = [r(m -$

1)/(I − 1)] of the blocks. This model can be written as a special case of (6.6.1) by considering the set of Ir equations

$$Y_{ijk} = \mu + A_i + B_j + E_{ijk}$$
$$i = 1, \ldots, I; \qquad j = 1, \ldots, J; \qquad k = K_{ij} \tag{6.7.1}$$

where μ, A_i, B_j, and E_{ijk} are defined in (6.6.1), $K_{ij} = 1$ if treatment i occurs in block j, and $K_{ij} = 0$, otherwise. The term Y_{ij0} is undefined.

Confidence intervals on σ_A^2 and σ_B^2 can be computed as discussed in Section 6.6.2. For this model, $E(S_{1W}^2) = \sigma_E^2 + [J(m-1)/(I-1)]\sigma_A^2 = \theta_{1W}$ (treatment adjusted), $E(S_{2W}^2) = \sigma_E^2 + [I(r-1)/(J-1)]\sigma_B^2 = \theta_{2W}$ (blocks adjusted), and $E(S_{PW}^2) = \sigma_E^2$. For the special case where $I = J$, $n_1 S_{1W}^2/\theta_{1W}$, $n_2 S_{2W}^2/\theta_{2W}$, and $n_P S_{PW}^2/\sigma_E^2$ are jointly independent chi-squared random variables. Thus, when $I = J$ the results of Chapter 3 for constructing confidence intervals on the variance components are directly applicable. An example of this special case is now presented.

Example 6.7.1 Graybill (1961, p. 313) describes a greenhouse experiment used to determine the effects of seven fertilizer treatments on nitrogen content of alfalfa forage. Seven blocks of size three plots were formed to account for location of plants within the greenhouse. For purposes of illustration, we analyze these data assuming both treatments and blocks are random. Table 6.7.1 reports the ANOVA table for the data. In the notation of this section, $I = 7$, $J = 7$, $r = 3$, $m = 3$, and $\lambda = 1$. Since $I = J$, $n_1 S_{1W}^2/\theta_{1W}$ and $n_P S_{PW}^2/\theta_{PW}$ are jointly independent chi-squared random variables and (3.3.1) and (3.3.2) can be used to construct a confidence interval on $\sigma_A^2 = (\theta_{1W} - \theta_{PW})/2.333$. To illustrate, using (3.3.2) a 97.5% upper bound on σ_A^2 is 1.031 where $H_1 = 3.849$, $G_P = .5438$, $H_{1P} = -.5233$, and $V_U = .8091$.

Table 6.7.1 Analysis of Variance for Balanced Incomplete Block

SV	DF	MS	EMS
Fertilizer (adj.)	$n_1 = 6$	$S_{1W}^2 = .5485$	$\theta_{1W} = \sigma_E^2 + 2.333\sigma_A^2$
Blocks (adj.)	$n_2 = 6$	$S_{2W}^2 = .2562$	$\theta_{2W} = \sigma_E^2 + 2.333\sigma_B^2$
Error	$n_P = 8$	$S_{PW}^2 = .2424$	$\theta_{PW} = \sigma_E^2$

Table 6.8.1 Summary Formulas for Confidence Intervals and MVU Estimators in the Two-Factor Balanced Crossed Design with Interaction

Parameter	MVU estimator	Confidence interval (approximate unless noted)
σ_E^2	S_4^2	(3.2.1)–Exact
$\sigma_A^2 \ [\sigma_B^2, \sigma_{AB}^2]$	$\dfrac{S_1^2 - S_3^2}{JK}\left[\dfrac{S_2^2 - S_3^2}{IK}, \dfrac{S_3^2 - S_4^2}{K}\right]$	(6.2.2)
$\gamma = \sigma_A^2 + \sigma_B^2 + \sigma_{AB}^2 + \sigma_E^2$	$\dfrac{IS_1^2 + JS_2^2 + (IJ - I - J)S_3^2 + IJ(K-1)S_4^2}{IJK}$	(3.2.5)
$\lambda_A = \dfrac{\sigma_A^2}{\sigma_E^2}\ \left[\lambda_B = \dfrac{\sigma_B^2}{\sigma_E^2}\right]$	$\dfrac{(S_1^2 - S_3^2)(1 - 2/n_4)}{JKS_4^2}\left[\dfrac{(S_2^2 - S_3^2)(1 - 2/n_4)}{IKS_4^2}\right]$	(6.2.3)
$\lambda_{AB} = \dfrac{\sigma_{AB}^2}{\sigma_E^2}$	$\dfrac{(S_3^2/S_4^2)(1 - 2/n_4) - 1}{K}$	(6.2.4)–Exact
$\lambda_A/\lambda_B = \sigma_A^2/\sigma_B^2$	None	(6.2.5)
$\rho_A = \dfrac{\sigma_A^2}{\gamma}$	None	(6.2.6)
$\rho_B = \dfrac{\sigma_B^2}{\gamma}$	None	(6.2.7)
$\rho_{AB} = \dfrac{\sigma_{AB}^2}{\gamma}$	None	(6.2.8)
$\rho_E = \dfrac{\sigma_E^2}{\gamma}$	None	(3.4.3),(3.4.4)

Table 6.8.2 Summary Formulas for Confidence Intervals and MVU Estimators in the Two-Factor Balanced Crossed Design Without Interaction

Parameter	MVU estimator	Confidence interval (approximate unless noted)
σ_E^2	S_P^2	(3.2.1)–Exact
σ_A^2 $[\sigma_B^2]$	$\dfrac{S_1^2 - S_P^2}{JK}\left[\dfrac{S_2^2 - S_P^2}{IK}\right]$	(6.2.2)-with substitutions
$\gamma = \sigma_A^2 + \sigma_B^2 + \sigma_E^2$	$\dfrac{IS_1^2 + JS_2^2 + (IJK - I - J)S_P^2}{IJK}$	(3.2.5)
$\lambda_A = \dfrac{\sigma_A^2}{\sigma_E^2}\left[\lambda_B = \dfrac{\sigma_B^2}{\sigma_E^2}\right]$	$\dfrac{(S_1^2/S_P^2)(1 - 2/n_P) - 1}{JK}\left[\dfrac{(S_2^2/S_P^2)(1 - 2/n_P) - 1}{IK}\right]$	(6.3.2)–Exact
$\lambda_A/\lambda_B = \sigma_A^2/\sigma_B^2$	None	(6.2.5)
$\rho_A = \dfrac{\sigma_A^2}{\gamma}$	None	(6.3.3)
$\rho_B = \dfrac{\sigma_B^2}{\gamma}$	None	(6.3.4)
$\rho_E = \dfrac{\sigma_E^2}{\gamma}$	None	(3.4.3),(3.4.4)

Broemeling and Bee (1976) used Kimball's inequality to construct simultaneous regions for σ_B^2/σ_E^2 and σ_A^2/σ_E^2 in a balanced incomplete block random model. However, by combining different sets of the sufficient statistics they define different regions. They make no recommendations on which sets are preferred. They note the results can be extended to partially balanced incomplete designs using the results of Weeks and Graybill (1962).

6.8 SUMMARY

Tables 6.8.1 and 6.8.2 summarize the results for the balanced designs presented in this chapter. The general approach for unbalanced designs is to apply the unweighted sums of squares in the balanced model equations. The unweighted sums of squares are defined in Table 6.5.2. When missing cells are present, the unweighted sums of squares cannot be defined. If the design is connected, then alternative sums of squares can be used in the same manner. Finally, the exact Wald interval discussed in Appendix B can be used for some ratios of interest. Computation of this interval is demonstrated in Examples 6.5.1, 6.6.1, and 6.6.2.

7
Mixed Models

7.1 INTRODUCTION

All models considered in previous chapters have contained only random effects. In this chapter we consider several mixed models that contain both fixed and random effects. Sections 7.2–7.6 consider balanced mixed models and Section 7.7 provides some results for unbalanced models. As in balanced random models, confidence intervals on variance components in balanced mixed models are based on independent mean squares from the ANOVA table. These mean squares have scaled chi-squared distributions and expected values that contain the variance components but not the fixed effects. Hence, confidence intervals on variance components in mixed models are formed in exactly the same manner as in random models.

In addition to confidence intervals on variance components, investigators employing mixed models are often interested in (1) tests of hypotheses concerning the fixed effects, (2) confidence intervals on estimable linear functions of the fixed effects, and (3) confidence intervals on linear combinations of estimable functions of the fixed effects and realized values of the random effects. We demonstrate how (1) and (2) are obtained for all of the balanced mixed models considered in this chapter. In general, exact tests with optimal properties exist for (1) in balanced mixed models whenever a proper mean square exists for the hypothesis of interest (see, e.g., Mathew and Sinha, 1988a). A proper mean square has a scaled chi-squared distribution with an expected value equal to the expected mean square of the fixed effect under the null hypothesis. When a proper mean square exists, exact intervals for (2) are also available. When a proper mean square does not exist, one can usually obtain a linear combination of mean squares that has an expected value equal to the expected mean square of the fixed effect under the null hypothesis. Several methods for obtaining (1) and (2) when such linear combinations exist have been proposed. We present some of these methods in Section 7.3 and demonstrate their application with a three-factor crossed mixed model with interaction. Readers interested in intervals of type (3) are referred to papers by Harville (1976), Gianola (1980), Jeyaratnam and Panchapakesan (1988), and Jeske and Harville (1988).

7.2 BALANCED TWO-FACTOR CROSSED MIXED MODELS

Probably the most familiar mixed linear model is the two-factor crossed model with interaction. This model has wide application in practice and has been the subject of much discussion concerning its formulation. One possible formulation of this model is

$$Y_{ijk} = \mu + \alpha_i + B_j + (\alpha B)_{ij} + E_{ijk}$$

$$\sum_i \alpha_i = 0 \tag{7.2.1}$$

$$i = 1, ..., I; \quad j = 1, ..., J; \quad k = 1, ..., K$$

where μ and α_i are fixed unknown constants, and B_j, $(\alpha B)_{ij}$, and E_{ijk} are mutually independent normal random variables with means of zero and variances σ_B^2, $\sigma_{\alpha B}^2$, and σ_E^2, respectively. The sources of variation and sums of squares for this model are the same ones reported in Table 6.2.1. The expected mean squares for model (7.2.1) are shown in Table 7.2.1. Note these are the same as those in Table 6.2.1 except σ_A^2 is replaced with $\sum_i \alpha_i^2/(I - 1)$. Under the assumptions of (7.2.1) the statistics $n_2 S_2^2/\theta_2$, $n_3 S_3^2/\theta_3$, and $n_4 S_4^2/\theta_4$ are jointly independent chi-

Table 7.2.1   Expected Mean Squares for Balanced Two-Factor Crossed Mixed Models with Interaction

SV	MS	EMS (7.2.1)	EMS (7.2.3)
A	S_1^2	$\theta_1 = \sigma_E^2 + K\sigma_{\alpha B}^2 + \dfrac{JK\sum_i \alpha_i^2}{I-1}$	$\theta_1 = \sigma_E^2 + \dfrac{IK\sigma_{\alpha B*}^2}{I-1} + \dfrac{JK\sum_i \alpha_i^2}{I-1}$
B	S_2^2	$\theta_2 = \sigma_E^2 + K\sigma_{\alpha B}^2 + IK\sigma_B^2$	$\theta_2 = \sigma_E^2 + IK\sigma_{B*}^2$
AB	S_3^2	$\theta_3 = \sigma_E^2 + K\sigma_{\alpha B}^2$	$\theta_3 = \sigma_E^2 + \dfrac{IK\sigma_{\alpha B*}^2}{I-1}$
Error	S_4^2	$\theta_4 = \sigma_E^2$	$\theta_4 = \sigma_E^2$

squared random variables. Thus, the confidence intervals for functions of σ_B^2, $\sigma_{\alpha B}^2$, and σ_E^2 are the same as those presented in Section 6.2.

The hypothesis concerning the fixed factor is $H_o : \alpha_i = 0$ for all i against H_a : at least one $\alpha_i \neq 0$. A proper mean square exists for this hypothesis since the expected value of S_1^2 when H_o is true is $\sigma_E^2 + K\sigma_{\alpha B}^2 = \theta_3$. Thus, an exact size α test of H_o is to reject H_o if $S_1^2/S_3^2 > F_{\alpha:n_1,n_3}$. This is the same test shown in Table 6.2.2 for σ_A^2.

Since a proper mean square exists for the fixed effect hypothesis, exact confidence intervals can be constructed on linear combinations of estimable functions of the α_i. In particular, if H_o is rejected, one typically is interested in contrasts of the form $\omega = \sum_i c_i \alpha_i$ where $\sum_i c_i = 0$. The best linear unbiased estimator for ω is $\hat{\omega} = \sum_i c_i \bar{Y}_{i**}$ where $\bar{Y}_{i**} = \sum_j\sum_k Y_{ijk}/(JK)$. Noting that $\mathrm{Var}(\bar{Y}_{i**}) = (\sigma_B^2 + \sigma_{\alpha B}^2 + \sigma_E^2/K)/J$ and $\mathrm{Cov}(\bar{Y}_{i**},\bar{Y}_{i'**}) = \sigma_B^2/J$, the variance of $\hat{\omega}$ is seen to be $\mathrm{Var}(\hat{\omega}) = (\sum_i c_i^2)(\sigma_E^2 + K\sigma_{\alpha B}^2)/(JK) = (\sum_i c_i^2)\theta_3/(JK)$. This variance is estimated by $\mathrm{var}(\hat{\omega}) = (\sum_i c_i^2)S_3^2/(JK)$. Note that $\mathrm{var}(\hat{\omega})$ is a function of the proper mean square used in the test for the fixed effect. Since $n_3 S_3^2/\theta_3$ has a chi-squared distribution with $n_3 = (I-1)(J-1)$ degrees of freedom and is independent of $\bar{Y}_{i**}$, an exact $1-\alpha$ confidence interval on ω is

$$\hat{\omega} \pm \sqrt{\dfrac{(\sum_i c_i^2)S_3^2 F_{\alpha:1,n_3}}{JK}} \qquad (7.2.2)$$

where S_3^2 is defined in Table 6.2.1.

An alternative formulation to model (7.2.1) places a restriction on the interaction terms. In particular, this model is written as

$$Y_{ijk} = \mu + \alpha_i + B_j^* + (\alpha B)_{ij}^* + E_{ijk} \qquad (7.2.3)$$

$$\sum_i \alpha_i = 0; \quad \sum_i (\alpha B)_{ij}^* = 0$$

$$i = 1, \dots, I; \quad j = 1, \dots, J; \quad k = 1, \dots, K$$

μ, α_i, and E_{ijk} are as defined in (7.2.1), B_j^* are independent normal random variables with means of zero and variance σ_{B*}^2, $(\alpha B)_{ij}^*$ are normal random variables with means of zero and variance $\sigma_{\alpha B*}^2$, and $B_j^*,(\alpha B)_{ij}^*$, and E_{ijk} are uncorrelated with each other. The constraint on the interaction terms is a carryover from similar restrictions placed on interaction effects in the fixed effects model. A consequence of this restriction is that the $(\alpha B)_{ij}^*$ are now correlated where $\text{Cov}[(\alpha B)_{ij}^*, (\alpha B)_{i'j}^*] = -\sigma_{\alpha B*}^2/(I - 1)$ for all j. Searle (1971, p. 404) compares (7.2.3) with (7.2.1) and shows $\sigma_{B*}^2 = \sigma_B^2 + \sigma_{\alpha B}^2 / I$, and $\sigma_{\alpha B*}^2 = \sigma_{\alpha B}^2 (1 - 1/I)$. The ANOVA table for model (7.2.3) has the same sources of variation and sums of squares as Table 6.2.1. Table 7.2.1 shows the expected mean squares for (7.2.3). As with model (7.2.1), $n_2 S_2^2/\theta_2$, $n_3 S_3^2/\theta_3$, and $n_4 S_4^2/\theta_4$ form a set of jointly independent chi-squared random variables. Thus, the results of Chapter 3 can be used for constructing intervals on functions of σ_{B*}^2, $\sigma_{\alpha B*}^2$, and σ_E^2. For example, using the expected mean squares shown in Table 7.2.1 we note $\sigma_{B*}^2 = (\theta_2 - \theta_4)/(IK)$ and $\sigma_{\alpha B*}^2 = (\theta_3 - \theta_4)((I - 1)/(IK))$. The results of Section 3.3.1 can then be applied to construct intervals on σ_{B*}^2 and $\sigma_{\alpha B*}^2$. The hypothesis test for the fixed effect and the confidence interval on $\omega = \Sigma_i c_i \alpha_i$ are the same as those for model (7.2.1).

Both (7.2.1) and (7.2.3) are special cases of a more general formulation of the two-factor mixed model proposed by Scheffé (1956, 1959, Chapter 8). We denote the variance components defined by Scheffé in this general model as σ_E^2, σ_{SB}^2, and σ_{SAB}^2. The component σ_E^2 is as defined in (7.2.1) and (7.2.3) and σ_{SB}^2 and σ_{SAB}^2 are defined in equations (8.1.10) and (8.1.11) of Scheffé (1959, p. 264). We note that the methods used for constructing intervals on the variance components in (7.2.1) and (7.2.3) can also be used to construct intervals on functions of σ_E^2 and σ_{SB}^2. However, they cannot be used for functions of σ_{SAB}^2 because the sum of squares SS_{AB} defined in (8.1.16) of Scheffé (1959) is not a scaled chi-squared random variable. Hocking (1973) explains the relationships among Scheffé's model, (7.2.1), and (7.2.3) and discusses considerations for selecting an appropriate model.

Smith and Murray (1984) proposed a two-factor mixed model formulation that employs covariance components to permit negative correlations among observations within the same cell. As illustrated with the one-fold nested design in Section 4.2.5, the methods of Chapter 3 can be used to construct confidence intervals on the covariance components of this model.

To conclude this section, we consider the balanced two-factor crossed mixed model without interaction. This model is often described as either a repeated measures design or a randomized complete block design with fixed treatments and random blocks. For this model, $\sigma_{\alpha B}^2 = \sigma_{\alpha B}^2* = 0$ and $\sigma_B^2 = \sigma_B^2*$. The ANOVA table for this model is exactly the same as Table 6.3.2 but with σ_A^2 replaced with $\Sigma_i \alpha_i^2/(I - 1)$ in the expected mean square column. Factor A denotes the fixed treatment and factor B the random block effect. The distributional properties of S_2^2 and S_P^2 are the same as in Table 6.3.2 and intervals on σ_B^2 and σ_E^2 are obtained from the results in Section 6.3. The hypothesis of no fixed effect is tested by rejecting the null hypothesis if $S_I^2 / S_P^2 > F_{\alpha:n_1,n_p}$. The exact $1 - \alpha$ confidence interval on $\omega = \Sigma_i c_i \alpha_i$ with $\Sigma_i c_i = 0$ is

$$\hat{\omega} \pm \sqrt{\frac{(\Sigma_i\ c_i^2)S_P^2 F_{\alpha:1,n_p}}{JK}} \qquad (7.2.4)$$

where $\omega = \Sigma_i c_i \bar{Y}_{i**}$ and n_p and S_P^2 are defined in Table 6.3.2.

Example 7.2.1 Table 7.2.2 reports an ANOVA table for a subset of the coded data reported by Johnson and Leone (1977, p. 761). The data were used to investigate factors that affect yield strength of a jet engine compressor rotor blade. The computations shown in Table 7.2.2 are based on the data in Table 14.38 of Johnson and Leone for Vendor A with a 1″ bar size. The two factors shown in the table are Test Specimen (A) and Heat (B). The factor Heat has three random levels and the factor Test Specimen has three fixed levels: raw bar stock (BS), forgedowns (FD), and finished forged blades (B). The sample means for these three levels are 125.67 (BS), 196.33 (FD), and 195.00 (B).

Table 7.2.2 Analysis of Variance for
Jet Engine Data

SV	DF	MS
Test specimen (A)	$n_1 = 2$	14704
Heat (B)	$n_2 = 2$	9093.8
Test specimen × heat	$n_3 = 4$	1254.1
Error	$n_4 = 18$	57.778

There are three replicates for each of the nine test specimen-heat combinations.

A test for the fixed test specimen effect using model (7.2.1), i.e., $H_o: \alpha_i = 0$ for all i against H_a: at least one $\alpha_i \neq 0$, is conducted using the ratio S_1^2/S_3^2. Using Table 7.2.2 we reject H_o since $14704/1254.1 = 11.7 > F_{.05;2,4} = 6.94$. An exact 97.5% confidence interval on the mean difference of B and BS, i.e. $\omega = \alpha_3 - \alpha_1$, is [10.98 ; 127.7]. This computed interval is based on (7.2.2) with $\hat{\omega} = 195.00 - 125.67 = 69.33$, $J = 3$, $K = 3$, $F_{.025;1,4} = 12.2179$, and $\Sigma_i c_i^2 = 2$. If model (7.2.3) is assumed instead of (7.2.1), these fixed effect computations remain the same. To illustrate how a confidence interval on a variance component is computed, consider a confidence interval on the heat component, σ_B^2, under model (7.2.1). From Table 7.2.1 $\sigma_B^2 = (\theta_2 - \theta_3)/(IK)$ and a confidence interval on σ_B^2 is formed using (3.3.1) and (3.3.2). In particular, a 90% confidence interval on σ_B^2 is [24.20 ; 19534] where $V_L = G_2^2 S_2^4 + H_3^2 S_3^4 + G_{23} S_2^2 S_3^2$ and $V_U = H_2^2 S_2^4 + G_3^2 S_3^4 + H_{23} S_2^2 S_3^2$. A 90% confidence interval on σ_B^2* under model (7.2.3) is [330.7 ; 19692]. This interval is based on (3.3.1) and (3.3.2) by noting $\sigma_B^2* = (\theta_2 - \theta_4)/(IK)$ so that $V_L = G_2^2 S_2^4 + H_4^2 S_4^4 + G_{24} S_2^2 S_4^2$ and $V_U = H_2^2 S_2^4 + G_4^2 S_4^4 + H_{24} S_2^2 S_4^2$.

Example 7.2.2 Results for a randomized complete block experiment with random blocks and fixed treatments are reported by Sokal and Rohlf (1969, p.327). The experiment measured dry weights (in mg) of $I = 3$ fixed genotypes of beetles reared in $J = 4$ random blocks of flour. The average weights for the genotypes are .9568 (++), .9845 (+b), and .9153 (bb). The ANOVA table for the data is shown in Table 7.2.3.

Table 7.2.3   Analysis of Variance for Beetle Data

SV	DF	MS
Genotypes	$n_1 = 2$	.004858
Series (blocks)	$n_2 = 3$	.007130
Error	$n_p = 6$	.000697

The presence of differences among the genotypes is tested using the statistic S_1^2/S_P^2. In particular we reject $H_o : \alpha_i = 0$ for all i at the .05 significance level since $S_1^2/S_P^2 = .004858/.000697 = 6.97 > F_{.05:2,6} = 5.14$. Interval (7.2.4) can be used for constructing confidence intervals on the pairwise differences among the genotype means. For example, a 97.5% confidence interval on $\alpha_2 - \alpha_3$ is [.0138; .1246] where $\hat{\omega} = .9845 - .9153 = .0692$, $J = 4$, $K = 1$, $\Sigma_i c_i^2 = 2$, and $F_{.025:1,6} = 8.8131$. Finally, a confidence interval on σ_B^2 is computed using (3.3.1) and (3.3.2) as illustrated in Example 7.2.1. For the data in Table 7.2.3 a 95% confidence interval on σ_B^2 is [.00038 ; .0328] where $V_L = G_2^2 S_2^4 + H_P^2 S_P^4 + G_{2P} S_2^2 S_P^2$, and $V_U = H_2^2 S_2^4 + G_P^2 S_P^4 + H_{2P} S_2^2 S_P^2$.

7.3   A BALANCED THREE-FACTOR CROSSED MIXED MODEL WITH INTERACTION

We now present a three-factor crossed mixed model with interaction in which factor A is fixed. Using a model formulation analogous to (7.2.1), we define this mixed model as

$$Y_{ijkm} = \mu + \alpha_i + B_j + C_k + (\alpha B)_{ij} + (\alpha C)_{ik} + (BC)_{jk} + (\alpha BC)_{ijk} + E_{ijkm}$$

$$\sum_i \alpha_i = 0 \qquad\qquad (7.3.1)$$

$$i = 1, ..., I; \qquad j = 1, ..., J; \qquad k = 1, ..., K; \qquad m = 1, ..., M$$

where μ and α_i are fixed unknown constants, B_j, C_k, $(\alpha B)_{ij}$, $(\alpha C)_{ik}$, $(BC)_{jk}$, $(\alpha BC)_{ijk}$, and E_{ijkm} are mutually independent normal random variables with means of zero and variances σ_B^2, σ_C^2, $\sigma_{\alpha B}^2$, $\sigma_{\alpha C}^2$, σ_{BC}^2,

$\sigma^2_{\alpha BC}$, and σ^2_E, respectively. The ANOVA table for model (7.3.1) is the same as Table 6.4.1 with σ^2_A, σ^2_{AB}, σ^2_{AC}, and σ^2_{ABC} replaced with $\Sigma_i \alpha_i^2/(I-1)$, $\sigma^2_{\alpha B}$, $\sigma^2_{\alpha C}$, and $\sigma^2_{\alpha BC}$, respectively. Imhof (1960) proposed a more general formulation of the three-factor crossed mixed model by extending the two-factor mixed model proposed by Scheffé. The distributional properties of S_2^2, ..., S_8^2 are the same under (7.3.1) as they are under (6.4.1) and hence intervals on functions of the variance components in (7.3.1) can be obtained from the intervals reported in Section 6.4.

No exact test for $H_o : \alpha_i = 0$ for all i against H_a : at least one $\alpha_i \neq 0$ can be performed using the mean squares shown in the ANOVA table. This is because the expected value of S_1^2 under H_o is $\sigma^2_E + M \sigma^2_{\alpha BC} + JM\sigma^2_{\alpha C} + KM\sigma^2_{\alpha B}$ and no other single mean square has this expectation. This was the same problem encountered in Section 6.4 for the test $H_o : \sigma^2_A = 0$. Since no proper mean square exists, no exact confidence interval on $\omega = \Sigma_i c_i \alpha_i$ is available from the ANOVA table.

An exact test for the fixed effect hypothesis has been proposed by Seifert (1979,1981) and exact simultaneous confidence intervals on estimable functions of the α_i have been developed by Khuri (1984). However, although these methods achieve exactness, they do not employ all of the information contained in the set of ANOVA mean squares. As an alternative, we prefer approximate methods that make use of all this information. In general, these methods maintain the stated test size and have greater power than the exact methods. We now describe two of these approximate methods.

A popular method for constructing an approximate test of $H_o : \alpha_i = 0$ for all i against H_a : at least one $\alpha_i \neq 0$ is based on the Satterthwaite approximation discussed in Section 3.2.2. To illustrate this approach, we note that when H_o is true, $E(S_1^2) = \sigma^2_E + M \sigma^2_{\alpha BC} + JM\sigma^2_{\alpha C} + KM\sigma^2_{\alpha B} = \theta_4 + \theta_5 - \theta_7 = \gamma$, and $n_1 S_1^2/\gamma$ has a chi-squared distribution with $n_1 = I - 1$ degrees of freedom. Although the estimator for γ, $\hat{\gamma} = S_4^2 + S_5^2 - S_7^2$, is independent of S_1^2, $\hat{\gamma}$ is not a scaled chi-squared random variable. However, we can approximate $m\hat{\gamma}/\gamma$ as a chi-squared random variable with m degrees of freedom where

$$m = \frac{\hat{\gamma}^2}{S_4^4/n_4 + S_5^4/n_5 + S_7^4/n_7} \qquad (7.3.2)$$

Given that S_1^2 and $\hat{\gamma}$ are independent, an approximate F-test for the fixed effect is to reject H_o if $S_1^2/\hat{\gamma} > F_{\alpha:n_1,m}$.

A problem with the Satterthwaite test is that S_7^2 has a negative coefficient in $\hat{\gamma}$. Such an approximation may result in a test size that is greater than the stated level. Another problem associated with the Satterthwaite test is that although γ is positive, $\hat{\gamma}$ can be negative. The probability of a negative value for $\hat{\gamma}$ is greatest for small values of I, J, and K when $\sigma_{\alpha B}^2$ and $\sigma_{\alpha C}^2$ are small relative to $\sigma_{\alpha BC}^2$ and σ_E^2.

An alternative test recommended by Naik (1974) overcomes both of these difficulties. This test is based on the notion that if $\sigma_{\alpha BC}^2$ and σ_E^2 are small, the estimator $S_4^2 + S_5^2$ will not be badly biased for γ. The proposed test is to reject H_o if $T > 1$ where

$$T = \frac{S_1^2}{F_{\alpha:n_1,n_4}S_4^2 + F_{\alpha:n_1,n_5}S_5^2} \qquad (7.3.3)$$

with $n_1 = I - 1$, $n_4 = (I - 1)(J - 1)$, and $n_5 = (I - 1)(K - 1)$. The test is exact when either (i) $\sigma_E^2 = \sigma_{\alpha BC}^2 = \sigma_{\alpha B}^2 = 0$, or (ii) $\sigma_E^2 = \sigma_{\alpha BC}^2 = \sigma_{\alpha C}^2 = 0$. Naik shows that for levels of significance typically used in practice, the size of the test based on (7.3.3) is never greater than the stated level. However, the test can often have a size much less than the stated level and this is a disadvantage compared to the Satterthwaite procedure. The test is particularly conservative if either σ_E^2 or $\sigma_{\alpha BC}^2$ is large relative to $\sigma_{\alpha B}^2$ and $\sigma_{\alpha C}^2$. Unfortunately, the Satterthwaite approximation cannot be recommended in this situation because it has a relatively high probability of producing a negative value for $\hat{\gamma}$.

Because no proper mean square exists for the fixed effect hypothesis, no exact interval similar to (7.2.2) is available for the linear contrast $\omega = \Sigma_i c_i \alpha_i$ where $\Sigma_i c_i = 0$. To see this, note that the best linear unbiased estimator, $\hat{\omega} = \Sigma_i c_i \bar{Y}_{i***}$, has variance $\text{Var}(\hat{\omega}) = (\Sigma_i c_i^2)(KM\sigma_{\alpha B}^2 + JM\sigma_{\alpha C}^2 + M\sigma_{\alpha BC}^2 + \sigma_E^2)/(JKM) = (\Sigma_i c_i^2)(\theta_4 + \theta_5 - \theta_7)/(JKM)$. Thus, $\text{Var}(\hat{\omega})$ cannot be estimated using only a single mean square. The overall tests recommended by Satterthwaite and Naik could be used to construct confidence intervals on ω. However, as previously noted, each of these methods has problems in certain situations. In particular, the Satterthwaite method can be very liberal or produce a negative value of $\hat{\gamma}$, and the Naik method can be too conservative.

Calvin, Jeyaratnam, and Graybill (1986) compared the Naik and Satterthwaite intervals on ω with two alternative approaches. One of these methods was based on an approximation by Banerjee (1960) and the other was based on an approximation by Milliken and Johnson (1984, p. 281). The study concluded that although all of the intervals worked well in certain situations, the interval based on Banerjee's approximation performed well under all the examined conditions. This $1 - \alpha$ approximate confidence interval on ω is

$$\hat{\omega} \pm \sqrt{\frac{(\Sigma c_i^2)\text{Max}[F_{\alpha:1,n_4}S_4^2 + F_{\alpha:1,n_5}S_5^2 - F_{\alpha:1,n_7}S_7^2; F_{\alpha:1,n_7}S_7^2]}{JKM}} \qquad (7.3.4)$$

To conclude this section, we note that if two of the effects in (7.3.1) are fixed, proper mean squares exist for all the fixed effect hypotheses. Thus, exact tests and confidence intervals can be obtained for estimable functions of the fixed effect parameters. The overall test for the random effect still has no proper mean square and should be analyzed as discussed in Section 6.4.

Example 7.3.1   Table 7.3.1 reports the ANOVA table for a data set reported by Montgomery (1984, p. 256). The data resulted from an experiment to examine factors affecting the pressure drop experienced across an expansion valve in a turbine. The three factors are gas temperature (A), turbine speed (B), and gas pressure (C). Temperature is a fixed factor and the other two factors are random. The response variable was coded before the computations in Table 7.3.1 were performed. The means for the fixed temperature levels are -2.63 (60°), 5.67 (75°), and -2.00 (90°). We use (7.3.3) to test for the significance of the temperature effect. Using $F_{.05:2,6} = 5.1433$, $F_{.05:2,4} = 6.9443$, $S_1^2 = 308.39$, $S_4^2 = 134.91$, and $S_5^2 = 44.77$, we compute $T = .307$. Since $T < 1$ we are unable to claim differences exist in the average pressure drop across the fixed temperature levels. A 97.5% confidence interval on the pairwise difference $\omega = \alpha_{75°} - \alpha_{60°}$ is $[-3.28 ; 19.9]$. This interval is computed using (7.3.4) with $\hat{\omega} = 8.3$, $\Sigma_i c_i^2 = 2$, $F_{.025:1,6} = 8.8131$, $F_{.025:1,4} = 12.2179$, $F_{.025:1,12} = 6.5538$, $J = 4$, $K = 3$, and $M = 2$. Confidence intervals on $\sigma_{\alpha B}^2$, $\sigma_{\alpha C}^2$, σ_{BC}^2, and $\sigma_{\alpha BC}^2$ are obtained using (3.3.1) and (3.3.2). For example, $\sigma_{\alpha B}^2 = (\theta_4 - \theta_7)/6$ and a 95% confidence interval is $[5.20 ; 106]$.

Table 7.3.1 Analysis of Variance for Pressure Drop Data

SV	DF	MS	EMS under (7.3.1)
Temperature (A)	$n_1 = 2$	308.39	$\theta_1 = \sigma_E^2 + 2\sigma_{\alpha BC}^2 + 8\sigma_{\alpha C}^2 + 6\sigma_{\alpha B}^2 + 24\dfrac{\sum_i \alpha_i^2}{2}$
Speed (B)	$n_2 = 3$	58.52	$\theta_2 = \sigma_E^2 + 2\sigma_{\alpha BC}^2 + 6\sigma_{\alpha B}^2 + 6\sigma_{BC}^2 + 18\sigma_B^2$
Pressure (C)	$n_3 = 2$	2.52	$\theta_3 = \sigma_E^2 + 2\sigma_{\alpha BC}^2 + 6\sigma_{BC}^2 + 8\sigma_{\alpha C}^2 + 24\sigma_C^2$
Temperature × speed	$n_4 = 6$	134.91	$\theta_4 = \sigma_E^2 + 2\sigma_{\alpha BC}^2 + 6\sigma_{\alpha B}^2$
Temperature × pressure	$n_5 = 4$	44.77	$\theta_5 = \sigma_E^2 + 2\sigma_{\alpha BC}^2 + 8\sigma_{\alpha C}^2$
Speed × pressure	$n_6 = 6$	40.37	$\theta_6 = \sigma_E^2 + 2\sigma_{\alpha BC}^2 + 6\sigma_{BC}^2$
Temperature × speed × pressure	$n_7 = 12$	19.26	$\theta_7 = \sigma_E^2 + 2\sigma_{\alpha BC}^2$
Error	$n_8 = 36$	34.67	$\theta_8 = \sigma_E^2$

7.4 BALANCED NESTED MIXED MODELS

We now consider nested models in which random effects are nested within fixed effects. A two-fold nested mixed model is represented as

$$Y_{ijk} = \mu + \alpha_i + B_{ij} + E_{ijk}$$
$$\sum_i \alpha_i = 0 \qquad (7.4.1)$$
$$i = 1, \ldots, I; \qquad j = 1, \ldots, J; \qquad k = 1, \ldots, K$$

where μ and α_i are fixed unknown constants, and B_{ij}, E_{ijk} are mutually independent normal random variables with means of zero and variances σ_B^2 and σ_E^2, respectively. The ANOVA table for (7.4.1) is the same as Table 5.2.1 with σ_A^2 replaced with $\sum_i \alpha_i^2/(I-1)$ in the expected mean square column. The statistics $n_2 S_2^2/\theta_2$ and $n_3 S_3^2/\theta_3$ are jointly independent chi-squared random variables under (7.4.1), and so the intervals on functions of σ_B^2 and σ_E^2 found in Section 5.2 are still valid.

The hypothesis $H_o : \alpha_i = 0$ for all i is rejected against H_a : at least one $\alpha_i \neq 0$ if $S_1^2/S_2^2 > F_{\alpha:n_1,n_2}$. Exact confidence intervals can be constructed on the linear combination $v = \sum_i c_i \mu_i$ where $\mu_i = \mu + \alpha_i$

and the c_i are any set of constants. The variance of $\hat{v} = \Sigma_i c_i \bar{Y}_{i**}$ is $\text{Var}(\hat{v}) = (\Sigma_i c_i^2)(\sigma_B^2/J + \sigma_E^2/(JK))$ which can be estimated by $(\Sigma_i c_i^2) S_2^2/(JK)$ where S_2^2 is defined in Table 5.2.1. An exact $1 - \alpha$ confidence interval on v is

$$\hat{v} \pm \sqrt{\frac{(\Sigma_i c_i^2) S_2^2 F_{\alpha:1,n_2}}{JK}} \qquad (7.4.2)$$

The results for the two-fold nested mixed model can be extended to other balanced mixed models in which random effects are nested within fixed effects. For example, consider the three-fold nested model

$$Y_{ijkm} = \mu + \alpha_i + B_{ij} + C_{ijk} + E_{ijkm}$$

$$\sum_i \alpha_i = 0 \qquad (7.4.3)$$

$$i = 1, \ldots, I; \quad j = 1, \ldots, J; \quad k = 1, \ldots, K; \quad m = 1, \ldots, M$$

where μ and α_i are fixed unknown constants, and B_{ij}, C_{ijk}, and E_{ijkm} are mutually independent normal random variables with means of zero and variances σ_B^2, σ_C^2, and σ_E^2, respectively. The ANOVA table for (7.4.3) is identical to Table 5.3.2 except that σ_A^2 is replaced with $\Sigma_i \alpha_i^2/(I - 1)$. Confidence intervals on the variance components are based on S_2^2, S_3^2, and S_4^2 as discussed in Section 5.3.

An exact $1 - \alpha$ confidence interval on $v = \Sigma_i c_i \mu_i$ with $\mu_i = \mu + \alpha_i$ is

$$\hat{v} \pm \sqrt{\frac{(\Sigma_i c_i^2) S_2^2 F_{\alpha:1,n_2}}{JKM}} \qquad (7.4.4)$$

where $\hat{v} = \Sigma_i c_i \bar{Y}_{i***}$, and n_2 and S_2^2 are defined in Table 5.3.2. Vidmar and Brunden (1980) and Marcuse (1949) derived formulas for selecting J, K, and M to minimize the cost of obtaining a sample.

Example 7.4.1 Steel and Torrie (1960, pp. 118–123) report the results of a study that examined the growth of mint plants. The experiment considered $I = 6$ fixed treatments representing combinations of the crossed factors temperature and hours of daylight. Each treatment was applied to a random sample of $J = 3$ pots, each containing a random sample of $K = 4$ mint plants. Thus, plants were nested within pots and pots were nested within the fixed treatments. Both pots and plants were random factors. The response variable was the one week stem growth of the mint plant. Tables 7.4.1 and 7.4.2 report the ANOVA table and the sample treatment means for the experiment.

Equation (5.2.4) is used to construct a confidence interval on σ_B^2. A 95% confidence interval is [.015 ; 1.23] where $G_2 = .4858$, $G_3 = .2913$, $H_2 = 1.725$, $H_3 = .5174$, $G_{23} = .0107$, and $H_{23} = -.0722$. Based on this confidence interval, it appears there is significant variability among the pots. The overall treatment effect is tested with the ratio $S_1^2/S_2^2 = 35.93/2.15 = 16.7$. Since this ratio exceeds $F_{.05:5,12} = 3.11$, the null hypothesis of no treatment effect is rejected. Equation (7.4.2) can now be used to examine contrasts among the cell means to determine the nature of the treatment differences. For example, a 97.5% confidence interval on the difference between low temperature-8 hours and high temperature-8 hours is [2.07 ; 5.13]. This interval is computed using $\hat{\nu} = 7.3 - 3.7 = 3.6$, $\Sigma_i c_i^2 = 2$, $F_{.025:1,12} = 6.5538$, $J = 3$, and $K = 4$.

Table 7.4.1   Analysis of Variance for Mint Plant Data

SV	DF		MS	EMS
Treatments (A)	$n_1 = 5$	35.93		$\theta_1 = \sigma_E^2 + 4\sigma_B^2 + 12\dfrac{\Sigma_i \alpha_i^2}{5}$
Among pots within treatments (B)	$n_2 = 12$	2.15		$\theta_2 = \sigma_E^2 + 4\sigma_B^2$
Among plants within pots	$n_3 = 54$	.93		$\theta_3 = \sigma_E^2$

Table 7.4.2   Cell Means for Mint Plant Data

Temperature	Hours of daylight		
	8	12	16
Low	3.7	4.1	5.2
High	7.3	6.5	7.9

7.5 BALANCED MIXED MODELS WITH BOTH CROSSED AND NESTED EFFECTS: A SPLIT-PLOT DESIGN

The general methods of Chapter 3 can also be used in mixed models that contain both crossed and nested effects. To illustrate, we present a split-plot experiment in which a randomized complete block design is used for the whole plot experimental units. We assume the whole plot and subplot treatment effects are fixed and that blocks are random. This model is represented as

$$Y_{ijk} = \mu + R_i + \alpha_j + W_{ij} + \beta_k + (\alpha\beta)_{jk} + E_{ijk}$$

$$\sum_j \alpha_j = \sum_k \beta_k = \sum_j (\alpha\beta)_{jk} = \sum_k (\alpha\beta)_{jk} = 0 \qquad (7.5.1)$$

$$i = 1, ..., I; \qquad j = 1, ..., J; \qquad k = 1, ..., K$$

where μ, α_j, β_k, and $(\alpha\beta)_{jk}$ are fixed unknown constants, and R_i, W_{ij}, and E_{ijk} are mutually independent normal random variables with means of zero and variances σ_R^2, σ_W^2, and σ_E^2, respectively. The ANOVA table for model (7.5.1) is shown in Table 7.5.1 where $SS1 = JK\Sigma_i(\bar{Y}_{i**} - \bar{Y}_{***})^2$, $SS2 = IK\Sigma_j(\bar{Y}_{*j*} - \bar{Y}_{***})^2$, $SS3 =$

Table 7.5.1 Analysis of Variance for Split-Plot Randomized Complete Block Design

SV	DF	SS	MS	EMS
Blocks (Reps)	$n_1 = I - 1$	SS1	S_1^2	$\theta_1 = \sigma_E^2 + K\sigma_W^2 + JK\sigma_R^2$
A	$n_2 = J - 1$	SS2	S_2^2	$\theta_2 = \sigma_E^2 + K\sigma_W^2 + \dfrac{IK \sum_j \alpha_j^2}{J - 1}$
Whole plot error	$n_3 = (I - 1)(J - 1)$	SS3	S_3^2	$\theta_3 = \sigma_E^2 + K\sigma_W^2$
B	$n_4 = K - 1$	SS4	S_4^2	$\theta_4 = \sigma_E^2 + \dfrac{IJ \sum_k \beta_k^2}{K - 1}$
AB	$n_5 = (J - 1)(K - 1)$	SS5	S_5^2	$\theta_5 = \sigma_E^2 + \dfrac{I \sum_j \sum_k (\alpha\beta)_{jk}}{(J - 1)(K - 1)}$
Subplot error	$n_6 = J(K - 1)(I - 1)$	SS6	S_6^2	$\theta_6 = \sigma_E^2$
Total	$IJK - 1$			

$K\Sigma_i\Sigma_j(\bar{Y}_{ij*} - \bar{Y}_{i**} - \bar{Y}_{*j*} + \bar{Y}_{***})^2 = K\Sigma_i\Sigma_j(\bar{Y}_{ij*} - \bar{Y}_{***})^2 - SS1 - SS2$, $SS4 = IJ\Sigma_k(\bar{Y}_{**k} - \bar{Y}_{***})^2$, $SS5 = I\Sigma_j\Sigma_k(\bar{Y}_{*jk} - \bar{Y}_{***})^2 - SS2 - SS4$, and $SS6 = \Sigma_i\Sigma_j\Sigma_k(Y_{ijk} - \bar{Y}_{*jk} - \bar{Y}_{ij*} + \bar{Y}_{*j*})^2 = \Sigma_i\Sigma_j\Sigma_k(Y_{ijk} - \bar{Y}_{***})^2 - SS1 - SS2 - SS3 - SS4 - SS5$. Since the random effects in model (7.5.1) are nested, S_1^2, S_3^2, and S_6^2 have the same expected mean squares as S_1^2, S_2^2, and S_3^2, respectively, in Table 5.2.1. Thus, confidence intervals on functions of σ_R^2, σ_W^2, and σ_E^2 are obtained from the results of Section 5.2.

Hypothesis tests for the fixed effect parameters in (7.5.1) are shown in Table 7.5.2. All of these tests are exact and have optimal properties. Confidence intervals on the fixed effect parameters are obtained in the following manner. The best linear unbiased estimator for the contrast $\omega_\alpha = \Sigma_j c_j \alpha_j$ with $\Sigma_j c_j = 0$ is $\hat{\omega}_\alpha = \Sigma_j c_j \bar{Y}_{*j*}$. The variance of $\hat{\omega}_\alpha$ is $Var(\hat{\omega}_\alpha) = (\Sigma_j c_j^2)(\sigma_E^2 + K\sigma_W^2)/(IK) = (\Sigma_j c_j^2)\theta_3/(IK)$. Using S_3^2 to estimate θ_3, an exact $1 - \alpha$ confidence interval on ω_α is

$$\hat{\omega}_\alpha \pm \sqrt{\frac{(\Sigma_j c_j^2)S_3^2 F_{\alpha:1,n_3}}{IK}} \qquad (7.5.2)$$

Table 7.5.2   Exact Size α Tests for Fixed Effect Hypotheses in Split-Plot Design

Hypotheses	Decision rule: Reject H_o if
$H_o : \alpha_j = 0$ for all j $H_a : \alpha_j \neq 0$ for at least one j	$\dfrac{S_2^2}{S_3^2} > F_{\alpha:n_2,n_3}$
$H_o : \beta_k = 0$ for all k $H_a : \beta_k \neq 0$ for at least one k	$\dfrac{S_4^2}{S_6^2} > F_{\alpha:n_4,n_6}$
$H_o : (\alpha\beta)_{jk} = 0$ for all j, k $H_a : (\alpha\beta)_{jk} \neq 0$ for at least one j, k	$\dfrac{S_5^2}{S_6^2} > F_{\alpha:n_5,n_6}$

In a similar manner, an exact $1 - \alpha$ confidence interval on the contrast $\omega_\beta = \Sigma_k c_k \beta_k$ with $\Sigma_k c_k = 0$ is

$$\hat{\omega}_\beta \pm \sqrt{\frac{(\Sigma_k c_k^2) S_6^2 F_{\alpha:1,n_6}}{IJ}} \qquad (7.5.3)$$

where $\hat{\omega}_\beta = \Sigma_k c_k \bar{Y}_{**k}$.

Differences of the cell means $\bar{Y}_{*jk} - \bar{Y}_{*j'k'}$ are often of interest in situations where the $(\alpha\beta)_{jk}$ interaction effect is significant. If $j = j'$, an exact $1 - \alpha$ confidence interval on the expected value of this difference is

$$\bar{Y}_{*jk} - \bar{Y}_{*jk'} \pm \sqrt{\frac{2 S_6^2 F_{\alpha:1,n_6}}{I}} \qquad (7.5.4)$$

If $j \neq j'$, $\text{Var}(\bar{Y}_{*jk} - \bar{Y}_{*j'k'}) = 2(\sigma_W^2 + \sigma_E^2)/I$. Examination of Table 7.5.1 shows that no single mean square can be used to provide an unbiased esimator of this variance. Thus, as was the case in Section 7.3, no exact confidence interval based on the ANOVA mean squares is available. Using the results of Burdick and Sielken (1978), an exact confidence interval on the expected difference of the cell means can be obtained. However, like the exact method of Khuri (1984), it suffers a loss of power in comparison to competing approximate methods. The Satterthwaite approximation is one such method. Here we approximate $m\hat{\gamma}/\gamma$ as a chi-squared random variable with m degrees of freedom where

$$\gamma = \sigma_W^2 + \sigma_E^2 \quad ; \quad \hat{\gamma} = \frac{S_3^2 + (K - 1)S_6^2}{K} \qquad \text{and}$$

$$m = \frac{\hat{\gamma}^2}{\dfrac{S_3^4}{K^2 n_3} + \dfrac{(K - 1)^2 S_6^4}{K^2 n_6}} \qquad (7.5.5)$$

Using (7.5.5), an approximate $1 - \alpha$ confidence interval on the expected value of $\bar{Y}_{*jk} - \bar{Y}_{*j'k'}$ for $j \neq j'$ is

$$\bar{Y}_{*jk} - \bar{Y}_{*j'k'} \pm \sqrt{\frac{2\hat{\gamma} F_{\alpha:1,m}}{I}} \qquad (7.5.6)$$

Table 7.5.3   Analysis of Variance for Seed Data

SV	DF	MS
Blocks (reps)	$n_1 = 3$	947.62
Seed lots (A)	$n_2 = 3$	949.34
Whole plot error	$n_3 = 9$	68.70
Seed treatments (B)	$n_4 = 3$	56.84
Interaction (AB)	$n_5 = 9$	65.16
Subplot error	$n_6 = 36$	20.31

In practice, the greatest integer less than m is used in determining $F_{\alpha:1,m}$. Unlike the Satterthwaite approximation employed in (7.3.2), $\hat{\gamma}$ has positive coefficients on both S_3^2 and S_6^2 and is always positive. Thus, the approximation is expected to work well unless n_3 is small and $n_6 - n_3$ is large. Milliken and Johnson (1984, p. 303) recommend using F^* in place of $F_{\alpha:1,m}$ in (7.5.6) where

$$F^* = \left(\frac{\sqrt{F_{\alpha:1,n_3}}S_3^2 + \sqrt{F_{\alpha:1,n_6}}(K-1)S_6^2}{S_3^2 + (K-1)S_6^2} \right)^2. \qquad (7.5.7)$$

Example 7.5.1   Steel and Torrie (1960, pp. 236–240) report the results of a split plot design used to study the yields of $J = 4$ lots of oats for $K = 4$ chemical seed treatments. The seed lots, factor A, were assigned at random to whole plots and the chemical seed treatments, factor B, were assigned at random to the subplots within each whole plot. Each seed lot was applied in $I = 4$ blocks. Tables 7.5.3 and 7.5.4 report the ANOVA table and the sample means for the 16 seed lot-seed treatment combinations.

Ignoring the differences in subscripts, a confidence interval on $\sigma_W^2 = (\theta_3 - \theta_6)/K$ is obtained from (5.2.4). In particular, a 95% confidence interval on σ_W^2 is [2.34 ; 52.0], where $S_3^2 = 68.70$, $S_6^2 = 20.31$, $K = 4$, $G_3 = .5269$, $G_6 = .3387$, $G_{36} = .0121$, $H_3 = 2.333$, $H_6 = .6873$, and $H_{36} = -.1409$. A contrast of interest concerning the fixed effects parameters is the difference between the average of the

Table 7.5.4 Cell Means for Seed Data

Seed lots	Check	Ceresan M	Panogen	Agrox	Seed lot means
Vicland (1)	36.1	50.6	45.9	37.3	42.5
Vicland (2)	50.9	55.4	53.1	54.3	53.4
Clinton	53.9	51.4	55.9	56.1	54.3
Branch	61.9	63.4	57.7	61.3	61.1
Seed treatment means	50.7	55.2	53.1	52.2	52.8

check seed with the other three seed treatments. This contrast is written as $\omega_\beta = 3\beta_1 - \beta_2 - \beta_3 - \beta_4$. Equation (7.5.3) is used to construct a 95% confidence interval on ω_b. The computed interval is $[-16.3\;;\;-.485]$ with $\hat{\omega}_\beta = -8.4$, $\Sigma_k c_k^2 = 12$, $S_6^2 = 20.31$, $I = 4$, $J = 4$, and $F_{.05:1,36} = 4.1132$. Since both endpoints of this interval are negative, it is concluded the average of the three seed treatment means is greater than the check mean. To illustrate how differences in cell means are examined, consider the comparison of Vicland(1)-Check with Vicland(1)-Agrox. In our notation $\bar{Y}_{*11} = 36.1$ and $\bar{Y}_{*14} = 37.3$. Since $j = j' = 1$, interval (7.5.4) is used to construct the 97.5% confidence interval on the expected value of $\bar{Y}_{*11} - \bar{Y}_{*14}$. This interval is $[-8.65\;;\;6.25]$. Finally, consider a pairwise difference such as $\bar{Y}_{*11} - \bar{Y}_{*24} = 36.1 - 54.3$ where $j \neq j'$. A 97.5% confidence interval on the expected value of this difference given by (7.5.6) is $[-27.8\;;\;-8.62]$ where $\hat{\gamma} = 32.4$, $K = 4$, $m = 26.8$ (truncated to 26), $F_{.025:1,26} = 5.6586$, and $I = 4$.

7.6 REGRESSION MODELS WITH NESTED ERROR STRUCTURE

In many applications of regression analysis it is of interest to quantify the measurement error associated with the response variable. Consider the regression model

$$Y_i^* = \beta_o + \beta_1 X_i + E_i \qquad i = 1, ..., I \qquad (7.6.1)$$

where Y_i^* represents the true value of the response variable associated with X_i, β_o and β_1 are the fixed regression parameters, X_i is a known constant, and E_i is a normal random variable with a mean of zero and variance σ_E^2. Assume that J measurements of the response variable are made on each of the I experimental units. These measurements can be represented for the ith experimental unit as

$$Y_{ij} = Y_i^* + M_{ij} \qquad j = 1, \ldots, J \qquad (7.6.2)$$

where M_{ij} is a normal random variable with a mean of zero and variance σ_M^2. The random variable M_{ij} is viewed as the measurement error associated with the response variable. By combining (7.6.1) and (7.6.2) one obtains

$$Y_{ij} = \beta_o + \beta_1 X_i + E_i + M_{ij} \qquad (7.6.3)$$

Model (7.6.3) is a mixed model in which one desires confidence intervals on functions of σ_E^2, σ_M^2, and β_1. The ANOVA table for model (7.6.3) is shown in Table 7.6.1 where

$$\hat{\beta}_1 = \frac{\sum_i (X_i - \bar{X}_*)(\bar{Y}_{i*} - \bar{Y}_{**})}{\sum_i (X_i - \bar{X}_*)^2} \qquad (7.6.4)$$

$\bar{Y}_{i*} = \Sigma_j Y_{ij}/J$, $\bar{Y}_{**} = \Sigma_i \Sigma_j Y_{ij}/(IJ)$, $\bar{X}_* = \Sigma_i X_i/I$, $SS1 = J\hat{\beta}_1^2 \Sigma_i (X_i - \bar{X}_*)^2$, $SS2 = J\Sigma_i (\bar{Y}_{i*} - \bar{Y}_{**})^2 - SS1$, and

Table 7.6.1 Analysis of Variance for Balanced Regression Model with Nested Error Structure

SV	DF	SS	MS	EMS
Regression	$n_1 = 1$	SS1	S_1^2	$\theta_1 = \sigma_M^2 + J\sigma_E^2 + J\beta_1^2 \Sigma_i (X_i - \bar{X}_*)^2$
Error in equation	$n_2 = I - 2$	SS2	S_2^2	$\theta_2 = \sigma_M^2 + J\sigma_E^2$
Measurement error	$n_3 = I(J - 1)$	SS3	S_3^2	$\theta_3 = \sigma_M^2$
Total	$IJ - 1$			

$SS3 = \Sigma_i\Sigma_j(Y_{ij} - \bar{Y}_{i*})^2$. The statistics $n_2 S_2^2/\theta_2$ and $n_3 S_3^2/\theta_3$ are independent chi-squared random variables and can be used to form confidence intervals on functions of σ_E^2 and σ_M^2. In particular, results of Chapter 3 can be used by noting $\sigma_M^2 = \theta_3$, $\sigma_E^2 = (\theta_2 - \theta_3)/J$, $\sigma_E^2/\sigma_M^2 = (\theta_2/\theta_3 - 1)/J$, and $\sigma_E^2 + \sigma_M^2 = (\theta_2 + (J - 1)\theta_3)/J$.

The hypothesis $H_o : \beta_1 = 0$ is rejected in favor of $H_a : \beta_1 \neq 0$ if $S_1^2/S_2^2 > F_{\alpha:n_1,n_2}$. The variance of $\hat{\beta}_1$ is $\mathrm{Var}(\hat{\beta}_1) = (\sigma_E^2 + \sigma_M^2/J)/\Sigma_i(X_i - \bar{X}_*)^2$ which is estimated by $S_2^2/(J\Sigma_i(X_i - \bar{X}_*)^2)$. Since S_2^2 is independent of $\hat{\beta}_1$, an exact $1 - \alpha$ confidence interval on β_1 is

$$\hat{\beta}_1 \pm \sqrt{\frac{S_2^2 F_{\alpha:1,n_2}}{J\Sigma_i(X_i - \bar{X}_*)^2}} \qquad (7.6.5)$$

Example 7.6.1 Brownlee (1953, p. 69) reports the data shown in Table 7.6.2 where x represents the temperature °C of a certain process and y represents two repeated measurements of the yield of the process. Table 7.6.3 reports the ANOVA table based on these data and model (7.6.3). Using these results with (7.6.5) a 97.5% confidence interval on β_1 is $[-.067 ; .200]$ where $\beta_1 = .0665$, $F_{.025:1,2} = 38.5063$, $J = 2$, and $\Sigma_i(X_i - \bar{X}_*)^2 = 500$. A 95% confidence interval on the

Table 7.6.2 Temperature Data

x	10	20	30	40
y	92.8	94.0	95.1	94.9
	93.0	94.3	94.8	94.8

Table 7.6.3 ANOVA for Temperature Data

SV	DF	MS
Regression	$n_1 = 1$	4.4223
Error in equation	$n_2 = 2$	.46575
Measurement error	$n_3 = 4$	.02875

measurement error variance, σ_M^2, obtained from (2.6.1) is [.02875/ 2.7858 ; .02875/.1211] = [.0103 ; .237].

7.7 UNBALANCED MIXED MODELS

In this section we illustrate how the general strategy recommended in this book can be applied to unbalanced mixed models. We consider only the problem of constructing confidence intervals on functions of variance components in these models. Unlike the situation encountered with the balanced mixed model, tests of hypotheses and confidence intervals on estimable functions of fixed effects parameters are not simple to develop. This is because in unbalanced mixed models, best linear unbiased esimators of the fixed effects require that values of the variance components be known. When the variance components are unknown, they must be estimated, and several approaches for selecting point estimators must be considered. As stated in Chapters 1 and 2, there is a large amount of research on point estimation present in the literature. A good description of the various approaches used in un- balanced mixed models is provided by Searle (1987, Chapter 13) and Searle et al. (1992). Milliken and Johnson (1984, Chapter 23) also provide some examples of this analysis.

The approach used to construct confidence intervals on variance components in unbalanced mixed models is the same one used for unbalanced random models. When a design is unbalanced, there is generally no unique set of sums of squares that meets the independence and chi-squaredness assumptions required for the methods in Chapter 3. However, as illustrated with the random model, unweighted sums of squares work well in these equations, and can be recommended in even very unbalanced designs. We now provide two examples to illustrate this general approach.

Example 7.7.1 Consider an unbalanced two-fold nested mixed model represented as

$$Y_{ijk} = \mu + \alpha_i + B_{ij} + E_{ijk}$$

$$\sum_i \alpha_i = 0 \qquad\qquad (7.71)$$

$$i = 1, ..., I; \qquad j = 1, ..., J_i; \qquad k = 1, ..., K_{ij}$$

where the fixed and random effects are defined in (7.4.1). Molińska and Moliński (1989) report results of an experiment that examined the influence of parentage on the content of milk fat. Table 7.7.1 reports the data collected in the experiment. The factor male (A) is fixed and the nested factor herd (B) is random. Using the notation of Section 5.4, $S_{2U}^2 = S_2^2$ since $K_{ij} = 2$ for all i and j. Additionally, $n_2 S_2^2/\theta_2$ and $n_3 S_3^2/\theta_3$ are jointly independent chi-squared random variables. Thus, if we consider only the random effects, this design is balanced. The intervals on functions of σ_B^2 and σ_E^2 reported in Section 5.2 are therefore valid. Assume the values noted with asterisks (*) in Table 7.7.1 are missing. The model is now unbalanced with respect to the random effects and results in Section 5.4 should be applied. To illustrate, after deleting the (*) values in Table 7.7.1, $n_2 = 8$, $n_3 = 9$, $S_{2U}^2 = 222.459$, $S_3^2 = 46.294$, and $w_{3U} = 1.6271$. A 90% confidence interval on σ_B^2 based on (5.4.3) is [29.1 ; 371]. This interval suggests that variability among herds contributes to the variability in milk fat. As illustrated in Example 5.4.3, an exact interval on $\lambda_B = \sigma_B^2/\sigma_E^2$ can be obtained using Wald's method. A conservative interval on λ_B is defined in (5.4.9). The exact 90% confidence interval on λ_B for these data is [.361 ; 9.44].

Table 7.7.1 Values of Milk Fat (in kg)

Male	Herd	Milk fat of daughter	
1	1	123.2	138.8
	2	125.9	138.9
	3	119.2	122.5*
2	1	160.3	171.1
	2	158.5	162.3*
3	1	172.3	180.4
	2	212.5	202.8
	3	214.5	208.2
	4	183.5	192.6
4	1	152.3	148.0*
	2	162.4	158.1
	3	168.2	170.1

*See Example 7.7.1.

Example 7.7.2 Milliken and Johnson (1984, p. 285) report the results of an unbalanced two-factor crossed mixed model resulting from an experiment to examine productivity scores in an industrial experiment. The data are shown in Table 7.7.2. The factor machine (A) is fixed and the factor person (B) is random. If we assume the model formulation in (7.2.1) with $k = 1, \ldots, K_{ij}$, the sums of squares associated with the random effects will have the properties discussed in Section 6.5. Thus, the results of Section 6.5 can be used for constructing confidence intervals on functions of σ_B^2, $\sigma_{\alpha B}^2$, and σ_E^2. To illustrate, the unweighted mean squares for the data in Table 7.7.2 are $S_{2U}^2 = 170.48$ and $S_{3U}^2 = 30.11$. Following the discussion of Section 6.5, an approximate confidence interval on σ_B^2 can be obtained from (6.2.2) by replacing S_1^2, S_3^2, J, and K with S_{2U}^2, S_{3U}^2, I, and $\bar{K}_H$, respectively. The resulting 90% confidence interval on σ_B^2 is [6.12 ; 114] where $I = 3$ and $\bar{K}_H = 2.0769$. An exact Wald interval on $\sigma_{\alpha B}^2 / \sigma_E^2$ can be obtained as described in Example 6.5.1. For the data in Table 7.7.2 this exact 90% confidence interval is [7.10 ; 44.9].

Gallo and Khuri (1990) proposed exact tests for the model considered in this example. These tests have the same disadvantages as those proposed by Khuri and Littell (1987) discussed in Section 6.5.

Table 7.7.2 Productivity Scores

| | Person | | | | | |
Machine	1	2	3	4	5	6
1	52.0	51.8	60.0	51.1	50.9	46.4
		52.8		52.3	51.8	44.8
					51.4	49.2
2	64.0	59.7	68.6	63.2	64.8	43.7
		60.0	65.8	62.8	65.0	44.2
		59.0		62.2		43.0
3	67.5	61.5	70.8	64.1	72.1	62.0
	67.2	61.7	70.6	66.2	72.0	61.4
	66.9	62.3	71.0	64.0	71.1	60.5

7.8 SUMMARY

This chapter has illustrated how the methods presented for random models in earlier chapters can be extended to mixed models. Additionally, some confidence intervals on fixed effects in balanced designs were presented. For unbalanced mixed models, the unweighted sums of squares are recommended for constructing confidence intervals on functions of the variance components. The Wald procedure described in Appendix B can be used in unbalanced mixed models to construct exact confidence intervals on some ratios of variance components.

APPENDIX A
F-Tables

Table A.1 *F*-values for Two-Sided 80% $(1 - 2\alpha)$ or One-Sided 90% $(1 - \alpha)$ Intervals—$F_{.90:r,c}$ and $F_{.10:r,c}$

r	c 1	2	3	4	5	6	7	8	9
1	0.0251	0.0202	0.0187	0.0179	0.0175	0.0172	0.0170	0.0168	0.0167
	39.8635	8.5263	5.5383	4.5448	4.0604	3.7760	3.5894	3.4579	3.3603
2	0.1173	0.1111	0.1091	0.1082	0.1076	0.1072	0.1070	0.1068	0.1066
	49.5000	9.0000	5.4624	4.3246	3.7797	3.4633	3.2574	3.1131	3.0065
3	0.1806	0.1831	0.1855	0.1872	0.1884	0.1892	0.1899	0.1904	0.1908
	53.5932	9.1618	5.3908	4.1909	3.6195	3.2888	3.0741	2.9238	2.8129
4	0.2200	0.2312	0.2386	0.2435	0.2469	0.2494	0.2513	0.2528	0.2541
	55.8330	9.2434	5.3426	4.1072	3.5202	3.1808	2.9605	2.8064	2.6927
5	0.2463	0.2646	0.2763	0.2841	0.2896	0.2937	0.2969	0.2995	0.3015
	57.2401	9.2926	5.3092	4.0506	3.4530	3.1075	2.8833	2.7264	2.6106
6	0.2648	0.2887	0.3041	0.3144	0.3218	0.3274	0.3317	0.3352	0.3381
	58.2044	9.3255	5.2847	4.0097	3.4045	3.0546	2.8274	2.6683	2.5509
7	0.2786	0.3070	0.3253	0.3378	0.3468	0.3537	0.3591	0.3634	0.3670
	58.9060	9.3491	5.2662	3.9790	3.3679	3.0145	2.7849	2.6241	2.5053
8	0.2892	0.3212	0.3420	0.3563	0.3668	0.3748	0.3811	0.3862	0.3904
	59.4390	9.3668	5.2517	3.9549	3.3393	2.9830	2.7516	2.5893	2.4694
9	0.2976	0.3326	0.3555	0.3714	0.3831	0.3920	0.3992	0.4050	0.4098
	59.8576	9.3805	5.2400	3.9357	3.3163	2.9577	2.7247	2.5612	2.4403
10	0.3044	0.3419	0.3666	0.3838	0.3966	0.4064	0.4143	0.4207	0.4260
	60.1950	9.3916	5.2304	3.9199	3.2974	2.9369	2.7025	2.5380	2.4163
12	0.3148	0.3563	0.3838	0.4032	0.4177	0.4290	0.4381	0.4455	0.4518
	60.7052	9.4081	5.2156	3.8955	3.2682	2.9047	2.6681	2.5020	2.3789
15	0.3254	0.3710	0.4016	0.4235	0.4399	0.4529	0.4634	0.4720	0.4793
	61.2203	9.4247	5.2003	3.8704	3.2380	2.8712	2.6322	2.4642	2.3396
20	0.3362	0.3862	0.4202	0.4447	0.4633	0.4782	0.4903	0.5004	0.5089
	61.7403	9.4413	5.1845	3.8443	3.2067	2.8363	2.5947	2.4246	2.2983
25	0.3427	0.3955	0.4316	0.4578	0.4780	0.4941	0.5073	0.5183	0.5278
	62.0545	9.4513	5.1747	3.8283	3.1873	2.8147	2.5714	2.3999	2.2725
30	0.3471	0.4018	0.4394	0.4668	0.4880	0.5050	0.5190	0.5308	0.5408
	62.2650	9.4579	5.1681	3.8174	3.1741	2.8000	2.5555	2.3830	2.2547
60	0.3583	0.4178	0.4593	0.4900	0.5140	0.5334	0.5496	0.5634	0.5754
	62.7943	9.4746	5.1512	3.7896	3.1402	2.7620	2.5142	2.3391	2.2085
120	0.3639	0.4260	0.4695	0.5019	0.5275	0.5483	0.5658	0.5807	0.5937
	63.0606	9.4829	5.1425	3.7753	3.1228	2.7423	2.4928	2.3162	2.1843
∞	0.3696	0.4343	0.4799	0.5142	0.5413	0.5637	0.5825	0.5987	0.6129
	63.3282	9.4912	5.1337	3.7607	3.1050	2.7222	2.4708	2.2926	2.1592

Table A.1 continued

r	c = 10	12	15	20	25	30	60	120	∞
1	0.0166	0.0165	0.0163	0.0162	0.0161	0.0161	0.0159	0.0159	0.0158
	3.2850	3.1765	3.0732	2.9747	2.9177	2.8807	2.7911	2.7478	2.7055
2	0.1065	0.1063	0.1061	0.1059	0.1058	0.1057	0.1055	0.1055	0.1054
	2.9245	2.8068	2.6952	2.5893	2.5283	2.4887	2.3933	2.3473	2.3026
3	0.1912	0.1917	0.1923	0.1929	0.1932	0.1935	0.1941	0.1945	0.1948
	2.7277	2.6055	2.4898	2.3801	2.3170	2.2761	2.1774	2.1300	2.0838
4	0.2551	0.2567	0.2584	0.2601	0.2612	0.2620	0.2639	0.2649	0.2659
	2.6053	2.4801	2.3614	2.2489	2.1842	2.1422	2.0410	1.9923	1.9449
5	0.3033	0.3060	0.3088	0.3119	0.3137	0.3151	0.3184	0.3202	0.3221
	2.5216	2.3940	2.2730	2.1582	2.0922	2.0492	1.9457	1.8959	1.8473
6	0.3405	0.3443	0.3483	0.3526	0.3553	0.3571	0.3621	0.3647	0.3674
	2.4606	2.3310	2.2081	2.0913	2.0241	1.9803	1.8747	1.8238	1.7741
7	0.3700	0.3748	0.3799	0.3854	0.3889	0.3913	0.3977	0.4012	0.4047
	2.4140	2.2828	2.1582	2.0397	1.9714	1.9269	1.8194	1.7675	1.7167
8	0.3940	0.3997	0.4058	0.4124	0.4167	0.4196	0.4275	0.4317	0.4362
	2.3771	2.2446	2.1185	1.9985	1.9292	1.8841	1.7748	1.7220	1.6702
9	0.4139	0.4204	0.4274	0.4351	0.4401	0.4435	0.4528	0.4578	0.4631
	2.3473	2.2135	2.0862	1.9649	1.8947	1.8490	1.7380	1.6842	1.6315
10	0.4306	0.4378	0.4457	0.4544	0.4600	0.4639	0.4746	0.4804	0.4865
	2.3226	2.1878	2.0593	1.9367	1.8658	1.8195	1.7070	1.6524	1.5987
12	0.4571	0.4657	0.4751	0.4855	0.4923	0.4971	0.5103	0.5175	0.5253
	2.2841	2.1474	2.0171	1.8924	1.8200	1.7727	1.6574	1.6012	1.5458
15	0.4856	0.4958	0.5070	0.5197	0.5280	0.5340	0.5504	0.5597	0.5698
	2.2435	2.1049	1.9722	1.8449	1.7708	1.7223	1.6034	1.5450	1.4871
20	0.5163	0.5284	0.5420	0.5575	0.5678	0.5753	0.5964	0.6085	0.6221
	2.2007	2.0597	1.9243	1.7938	1.7175	1.6673	1.5435	1.4821	1.4206
25	0.5360	0.5494	0.5647	0.5822	0.5941	0.6028	0.6276	0.6422	0.6589
	2.1739	2.0312	1.8939	1.7611	1.6831	1.6316	1.5039	1.4399	1.3753
30	0.5496	0.5641	0.5806	0.5998	0.6129	0.6225	0.6504	0.6672	0.6866
	2.1554	2.0115	1.8728	1.7382	1.6589	1.6065	1.4755	1.4094	1.3419
60	0.5858	0.6033	0.6237	0.6479	0.6649	0.6777	0.7167	0.7421	0.7743
	2.1072	1.9597	1.8168	1.6768	1.5934	1.5376	1.3952	1.3203	1.2400
120	0.6052	0.6245	0.6472	0.6747	0.6945	0.7095	0.7574	0.7908	0.8385
	2.0818	1.9323	1.7867	1.6433	1.5570	1.4989	1.3476	1.2646	1.1686
∞	0.6255	0.6469	0.6724	0.7039	0.7271	0.7452	0.8065	0.8557	1.0000
	2.0554	1.9036	1.7551	1.6074	1.5176	1.4564	1.2915	1.1926	1.0000

Table A.2 *F*-values for Two-Sided 90% $(1 - 2\alpha)$ or One-Sided 95% $(1 - \alpha)$ Intervals—$F_{.95:r,c}$ and $F_{.05:r,c}$

r	c 1	2	3	4	5	6	7	8	9
1	6.2E-3	5.0E-3	4.6E-3	4.5E-3	4.3E-3	4.3E-3	4.2E-3	4.2E-3	4.2E-3
	161.4	18.5128	10.1280	7.7086	6.6079	5.9874	5.5915	5.3177	5.1174
2	0.0540	0.0526	0.0522	0.0520	0.0518	0.0517	0.0517	0.0516	0.0516
	199.5	19.0000	9.5521	6.9443	5.7861	5.1433	4.7374	4.4590	4.2565
3	0.0987	0.1047	0.1078	0.1097	0.1109	0.1118	0.1125	0.1131	0.1135
	215.7	19.1643	9.2766	6.5914	5.4095	4.7571	4.3468	4.0662	3.8625
4	0.1297	0.1440	0.1517	0.1565	0.1598	0.1623	0.1641	0.1655	0.1667
	224.6	19.2468	9.1172	6.3882	5.1922	4.5337	4.1203	3.8379	3.6331
5	0.1513	0.1728	0.1849	0.1926	0.1980	0.2020	0.2051	0.2075	0.2095
	230.2	19.2964	9.0135	6.2561	5.0503	4.3874	3.9715	3.6875	3.4817
6	0.1670	0.1944	0.2102	0.2206	0.2279	0.2334	0.2377	0.2411	0.2440
	234.0	19.3295	8.9406	6.1631	4.9503	4.2839	3.8660	3.5806	3.3738
7	0.1788	0.2111	0.2301	0.2427	0.2518	0.2587	0.2641	0.2684	0.2720
	236.8	19.3532	8.8867	6.0942	4.8759	4.2067	3.7870	3.5005	3.2927
8	0.1881	0.2243	0.2459	0.2606	0.2712	0.2793	0.2857	0.2909	0.2951
	238.9	19.3710	8.8452	6.0410	4.8183	4.1468	3.7257	3.4381	3.2296
9	0.1954	0.2349	0.2589	0.2752	0.2872	0.2964	0.3037	0.3096	0.3146
	240.5	19.3848	8.8123	5.9988	4.7725	4.0990	3.6767	3.3881	3.1789
10	0.2014	0.2437	0.2697	0.2875	0.3007	0.3108	0.3189	0.3256	0.3311
	241.9	19.3959	8.7855	5.9644	4.7351	4.0600	3.6365	3.3472	3.1373
12	0.2106	0.2574	0.2865	0.3068	0.3220	0.3338	0.3432	0.3511	0.3576
	243.9	19.4125	8.7446	5.9117	4.6777	3.9999	3.5747	3.2839	3.0729
15	0.2201	0.2716	0.3042	0.3273	0.3447	0.3584	0.3695	0.3787	0.3865
	245.9	19.4291	8.7029	5.8578	4.6188	3.9381	3.5107	3.2184	3.0061
20	0.2298	0.2863	0.3227	0.3489	0.3689	0.3848	0.3978	0.4087	0.4179
	248.0	19.4458	8.6602	5.8025	4.5581	3.8742	3.4445	3.1503	2.9365
25	0.2358	0.2954	0.3343	0.3625	0.3842	0.4015	0.4158	0.4279	0.4382
	249.3	19.4558	8.6341	5.7687	4.5209	3.8348	3.4036	3.1081	2.8932
30	0.2398	0.3016	0.3422	0.3718	0.3947	0.4131	0.4284	0.4413	0.4523
	250.1	19.4624	8.6166	5.7459	4.4957	3.8082	3.3758	3.0794	2.8637
60	0.2499	0.3174	0.3626	0.3960	0.4222	0.4436	0.4616	0.4769	0.4902
	252.2	19.4791	8.5720	5.6877	4.4314	3.7398	3.3043	3.0053	2.7872
120	0.2551	0.3255	0.3731	0.4086	0.4367	0.4598	0.4792	0.4959	0.5105
	253.3	19.4874	8.5494	5.6581	4.3985	3.7047	3.2674	2.9669	2.7475
∞	0.2603	0.3338	0.3839	0.4216	0.4517	0.4765	0.4976	0.5159	0.5319
	254.3	19.4957	8.5264	5.6281	4.3650	3.6689	3.2298	2.9276	2.7067

Table A.2   continued

r	10	12	15	20	25	30	60	120	∞
					c				
1	4.1E-3	4.1E-3	4.1E-3	4.0E-3	4.0E-3	4.0E-3	4.0E-3	3.9E-3	3.9E-3
	4.9646	4.7472	4.5431	4.3512	4.2417	4.1709	4.0012	3.9201	3.8415
2	0.0516	0.0515	0.0515	0.0514	0.0514	0.0514	0.0513	0.0513	0.0513
	4.1028	3.8853	3.6823	3.4928	3.3852	3.3158	3.1504	3.0718	2.9957
3	0.1138	0.1144	0.1149	0.1155	0.1158	0.1161	0.1167	0.1170	0.1173
	3.7083	3.4903	3.2874	3.0984	2.9912	2.9223	2.7581	2.6802	2.6049
4	0.1677	0.1692	0.1707	0.1723	0.1733	0.1740	0.1758	0.1767	0.1777
	3.4780	3.2592	3.0556	2.8661	2.7587	2.6896	2.5252	2.4472	2.3719
5	0.2112	0.2138	0.2165	0.2194	0.2212	0.2224	0.2257	0.2274	0.2291
	3.3258	3.1059	2.9013	2.7109	2.6030	2.5336	2.3683	2.2899	2.2141
6	0.2463	0.2500	0.2539	0.2581	0.2608	0.2626	0.2674	0.2699	0.2726
	3.2172	2.9961	2.7905	2.5990	2.4904	2.4205	2.2541	2.1750	2.0986
7	0.2750	0.2797	0.2848	0.2903	0.2938	0.2962	0.3026	0.3060	0.3096
	3.1355	2.9134	2.7066	2.5140	2.4047	2.3343	2.1665	2.0868	2.0096
8	0.2988	0.3045	0.3107	0.3174	0.3217	0.3247	0.3327	0.3370	0.3416
	3.0717	2.8486	2.6408	2.4471	2.3371	2.2662	2.0970	2.0164	1.9384
9	0.3187	0.3254	0.3327	0.3405	0.3456	0.3492	0.3588	0.3640	0.3695
	3.0204	2.7964	2.5876	2.3928	2.2821	2.2107	2.0401	1.9588	1.8799
10	0.3358	0.3433	0.3515	0.3605	0.3663	0.3704	0.3815	0.3876	0.3940
	2.9782	2.7534	2.5437	2.3479	2.2365	2.1646	1.9926	1.9105	1.8307
12	0.3632	0.3722	0.3821	0.3931	0.4004	0.4055	0.4194	0.4272	0.4355
	2.9130	2.6866	2.4753	2.2776	2.1649	2.0921	1.9174	1.8337	1.7522
15	0.3931	0.4040	0.4161	0.4296	0.4386	0.4451	0.4629	0.4730	0.4841
	2.8450	2.6169	2.4034	2.2033	2.0889	2.0148	1.8364	1.7505	1.6664
20	0.4259	0.4391	0.4539	0.4708	0.4822	0.4904	0.5138	0.5273	0.5425
	2.7740	2.5436	2.3275	2.1242	2.0075	1.9317	1.7480	1.6587	1.5705
25	0.4471	0.4619	0.4787	0.4981	0.5114	0.5211	0.5489	0.5655	0.5845
	2.7298	2.4977	2.2797	2.0739	1.9554	1.8782	1.6902	1.5980	1.5061
30	0.4620	0.4780	0.4963	0.5177	0.5324	0.5432	0.5749	0.5940	0.6164
	2.6996	2.4663	2.2468	2.0391	1.9192	1.8409	1.6491	1.5543	1.4591
60	0.5019	0.5215	0.5445	0.5721	0.5916	0.6064	0.6518	0.6815	0.7198
	2.6211	2.3842	2.1601	1.9464	1.8217	1.7396	1.5343	1.4290	1.3180
120	0.5234	0.5453	0.5713	0.6029	0.6258	0.6434	0.6998	0.7397	0.7975
	2.5801	2.3410	2.1141	1.8963	1.7684	1.6835	1.4673	1.3519	1.2214
∞	0.5462	0.5707	0.6001	0.6367	0.6640	0.6854	0.7587	0.8187	1.0000
	2.5379	2.2962	2.0658	1.8432	1.7110	1.6223	1.3893	1.2539	1.0000

Table A.3 F-values for Two-Sided 95% $(1 - 2\alpha)$ or One-Sided 97.5% $(1 - \alpha)$ Intervals—$F_{.975:r,c}$ and $F_{.025:r,c}$

r	1	2	3	4	c 5	6	7	8	9
1	1.5E-3 647.8	1.3E-3 38.5063	1.2E-3 17.4434	1.1E-3 12.2179	1.1E-3 10.0070	1.1E-3 8.8131	1.1E-3 8.0727	1.0E-3 7.5709	1.0E-3 7.2093
2	0.0260 799.5	0.0256 39.0000	0.0255 16.0441	0.0255 10.6491	0.0254 8.4336	0.0254 7.2599	0.0254 6.5415	0.0254 6.0595	0.0254 5.7147
3	0.0573 864.2	0.0623 39.1655	0.0648 15.4392	0.0662 9.9792	0.0672 7.7636	0.0679 6.5988	0.0684 5.8898	0.0688 5.4160	0.0691 5.0781
4	0.0818 899.6	0.0939 39.2484	0.1002 15.1010	0.1041 9.6045	0.1068 7.3879	0.1087 6.2272	0.1102 5.5226	0.1114 5.0526	0.1123 4.7181
5	0.0999 921.8	0.1186 39.2982	0.1288 14.8848	0.1354 9.3645	0.1399 7.1464	0.1433 5.9876	0.1459 5.2852	0.1480 4.8173	0.1497 4.4844
6	0.1135 937.1	0.1377 39.3315	0.1515 14.7347	0.1606 9.1973	0.1670 6.9777	0.1718 5.8198	0.1756 5.1186	0.1786 4.6517	0.1810 4.3197
7	0.1239 948.2	0.1529 39.3552	0.1698 14.6244	0.1811 9.0741	0.1892 6.8531	0.1954 5.6955	0.2002 4.9949	0.2041 4.5286	0.2073 4.1970
8	0.1321 956.7	0.1650 39.3730	0.1846 14.5399	0.1979 8.9796	0.2076 6.7572	0.2150 5.5996	0.2208 4.8993	0.2256 4.4333	0.2295 4.1020
9	0.1387 963.3	0.1750 39.3869	0.1969 14.4731	0.2120 8.9047	0.2230 6.6811	0.2315 5.5234	0.2383 4.8232	0.2438 4.3572	0.2484 4.0260
10	0.1442 968.6	0.1833 39.3980	0.2072 14.4189	0.2238 8.8439	0.2361 6.6192	0.2456 5.4613	0.2532 4.7611	0.2594 4.2951	0.2646 3.9639
12	0.1526 976.7	0.1962 39.4146	0.2235 14.3366	0.2426 8.7512	0.2570 6.5245	0.2682 5.3662	0.2773 4.6658	0.2848 4.1997	0.2910 3.8682
15	0.1613 984.9	0.2099 39.4313	0.2408 14.2527	0.2629 8.6565	0.2796 6.4277	0.2929 5.2687	0.3036 4.5678	0.3126 4.1012	0.3202 3.7694
20	0.1703 993.1	0.2242 39.4479	0.2592 14.1674	0.2845 8.5599	0.3040 6.3286	0.3197 5.1684	0.3325 4.4667	0.3433 3.9995	0.3525 3.6669
25	0.1759 998.1	0.2330 39.4579	0.2707 14.1155	0.2982 8.5010	0.3196 6.2679	0.3369 5.1069	0.3511 4.4045	0.3632 3.9367	0.3736 3.6035
30	0.1796 1001.4	0.2391 39.4646	0.2786 14.0805	0.3077 8.4613	0.3304 6.2269	0.3488 5.0652	0.3642 4.3624	0.3772 3.8940	0.3884 3.5604
60	0.1892 1009.8	0.2548 39.4812	0.2992 13.9921	0.3325 8.3604	0.3589 6.1225	0.3806 4.9589	0.3989 4.2544	0.4147 3.7844	0.4284 3.4493
120	0.1941 1014.0	0.2628 39.4896	0.3099 13.9473	0.3455 8.3092	0.3740 6.0693	0.3976 4.9044	0.4176 4.1989	0.4349 3.7279	0.4501 3.3918
∞	0.1990 1018.3	0.2711 39.4979	0.3209 13.9021	0.3590 8.2573	0.3896 6.0153	0.4152 4.8491	0.4372 4.1423	0.4562 3.6702	0.4731 3.3329

Table A.3   continued

| r | \multicolumn{9}{c}{c} |
	10	12	15	20	25	30	60	120	∞
1	1.0E-3	1.0E-3	1.0E-3	1.0E-3	1.0E-3	1.0E-3	9.9E-4	9.9E-4	9.8E-4
	6.9367	6.5538	6.1995	5.8715	5.6864	5.5675	5.2856	5.1523	5.0239
2	0.0254	0.0254	0.0254	0.0253	0.0253	0.0253	0.0253	0.0253	0.0253
	5.4564	5.0959	4.7651	4.4613	4.2909	4.1821	3.9253	3.8046	3.6889
3	0.0694	0.0698	0.0702	0.0706	0.0708	0.0710	0.0715	0.0717	0.0719
	4.8256	4.4742	4.1528	3.8587	3.6943	3.5894	3.3425	3.2269	3.1161
4	0.1131	0.1143	0.1155	0.1168	0.1176	0.1182	0.1196	0.1203	0.1211
	4.4683	4.1212	3.8043	3.5147	3.3530	3.2499	3.0077	2.8943	2.7858
5	0.1511	0.1533	0.1556	0.1580	0.1595	0.1606	0.1633	0.1648	0.1662
	4.2361	3.8911	3.5764	3.2891	3.1287	3.0265	2.7863	2.6740	2.5665
6	0.1831	0.1864	0.1898	0.1935	0.1958	0.1974	0.2017	0.2039	0.2062
	4.0721	3.7283	3.4147	3.1283	2.9686	2.8667	2.6274	2.5154	2.4082
7	0.2100	0.2143	0.2189	0.2239	0.2270	0.2292	0.2351	0.2382	0.2414
	3.9498	3.6065	3.2934	3.0074	2.8478	2.7460	2.5068	2.3948	2.2875
8	0.2328	0.2381	0.2438	0.2500	0.2540	0.2568	0.2642	0.2682	0.2725
	3.8549	3.5118	3.1987	2.9128	2.7531	2.6513	2.4117	2.2994	2.1918
9	0.2523	0.2585	0.2653	0.2727	0.2775	0.2809	0.2899	0.2948	0.3000
	3.7790	3.4358	3.1227	2.8365	2.6766	2.5746	2.3344	2.2217	2.1136
10	0.2690	0.2762	0.2840	0.2925	0.2981	0.3020	0.3127	0.3185	0.3247
	3.7168	3.3736	3.0602	2.7737	2.6135	2.5112	2.2702	2.1570	2.0483
12	0.2964	0.3051	0.3147	0.3254	0.3325	0.3375	0.3512	0.3588	0.3670
	3.6209	3.2773	2.9633	2.6758	2.5149	2.4120	2.1692	2.0548	1.9447
15	0.3268	0.3375	0.3494	0.3629	0.3718	0.3783	0.3962	0.4063	0.4175
	3.5217	3.1772	2.8621	2.5731	2.4110	2.3072	2.0613	1.9450	1.8326
20	0.3605	0.3737	0.3886	0.4058	0.4174	0.4258	0.4498	0.4638	0.4795
	3.4185	3.0728	2.7559	2.4645	2.3005	2.1952	1.9445	1.8249	1.7085
25	0.3826	0.3976	0.4148	0.4347	0.4484	0.4584	0.4874	0.5048	0.5248
	3.3546	3.0077	2.6894	2.3959	2.2303	2.1237	1.8687	1.7462	1.6259
30	0.3982	0.4146	0.4334	0.4555	0.4709	0.4822	0.5155	0.5358	0.5597
	3.3110	2.9633	2.6437	2.3486	2.1816	2.0739	1.8152	1.6899	1.5660
60	0.4405	0.4610	0.4851	0.5143	0.5351	0.5509	0.6000	0.6325	0.6747
	3.1984	2.8478	2.5242	2.2234	2.0516	1.9400	1.6668	1.5299	1.3883
120	0.4636	0.4867	0.5141	0.5480	0.5727	0.5917	0.6536	0.6980	0.7631
	3.1399	2.7874	2.4611	2.1562	1.9811	1.8664	1.5810	1.4327	1.2684
∞	0.4882	0.5142	0.5457	0.5853	0.6151	0.6386	0.7203	0.7884	1.0000
	3.0798	2.7249	2.3953	2.0853	1.9055	1.7867	1.4821	1.3104	1.0000

Table A.4 *F*-values for Two-Sided 98% $(1 - 2\alpha)$ or One-Sided 99% $(1 - \alpha)$ Intervals—$F_{.99:r,c}$ and $F_{.01:r,c}$

r	1	2	3	4	5	6	7	8	9
1	2.5E-4	2.0E-4	1.9E-4	1.8E-4	1.7E-4	1.7E-4	1.7E-4	1.7E-4	1.7E-4
	4052.2	98.5025	34.1162	21.1977	16.2582	13.7450	12.2464	11.2586	10.5614
2	0.0102	0.0101	0.0101	0.0101	0.0101	0.0101	0.0101	0.0101	0.0101
	4999.5	99.0000	30.8165	18.0000	13.2739	10.9248	9.5466	8.6491	8.0215
3	0.0293	0.0325	0.0339	0.0348	0.0354	0.0358	0.0361	0.0364	0.0366
	5403.4	99.1662	29.4567	16.6944	12.0600	9.7795	8.4513	7.5910	6.9919
4	0.0472	0.0556	0.0599	0.0626	0.0644	0.0658	0.0668	0.0676	0.0682
	5624.6	99.2494	28.7099	15.9770	11.3919	9.1483	7.8466	7.0061	6.4221
5	0.0615	0.0753	0.0829	0.0878	0.0912	0.0937	0.0956	0.0972	0.0984
	5763.6	99.2993	28.2371	15.5219	10.9670	8.7459	7.4604	6.6318	6.0569
6	0.0728	0.0915	0.1023	0.1093	0.1143	0.1181	0.1211	0.1234	0.1254
	5859.0	99.3326	27.9107	15.2069	10.6723	8.4661	7.1914	6.3707	5.8018
7	0.0817	0.1047	0.1183	0.1274	0.1340	0.1391	0.1430	0.1462	0.1488
	5928.4	99.3564	27.6717	14.9758	10.4555	8.2600	6.9928	6.1776	5.6129
8	0.0888	0.1156	0.1317	0.1427	0.1508	0.1570	0.1619	0.1659	0.1692
	5981.1	99.3742	27.4892	14.7989	10.2893	8.1017	6.8400	6.0289	5.4671
9	0.0947	0.1247	0.1430	0.1557	0.1651	0.1724	0.1782	0.1829	0.1869
	6022.5	99.3881	27.3452	14.6591	10.1578	7.9761	6.7188	5.9106	5.3511
10	0.0996	0.1323	0.1526	0.1668	0.1774	0.1857	0.1923	0.1978	0.2023
	6055.8	99.3992	27.2287	14.5459	10.0510	7.8741	6.6201	5.8143	5.2565
12	0.1072	0.1444	0.1680	0.1848	0.1975	0.2074	0.2155	0.2223	0.2279
	6106.3	99.4159	27.0518	14.3736	9.8883	7.7183	6.4691	5.6667	5.1114
15	0.1152	0.1573	0.1846	0.2044	0.2195	0.2316	0.2415	0.2497	0.2568
	6157.3	99.4325	26.8722	14.1982	9.7222	7.5590	6.3143	5.5151	4.9621
20	0.1235	0.1710	0.2025	0.2257	0.2437	0.2583	0.2704	0.2806	0.2893
	6208.7	99.4492	26.6898	14.0196	9.5526	7.3958	6.1554	5.3591	4.8080
25	0.1287	0.1796	0.2139	0.2394	0.2594	0.2757	0.2893	0.3008	0.3108
	6239.8	99.4592	26.5790	13.9109	9.4491	7.2960	6.0580	5.2631	4.7130
30	0.1322	0.1855	0.2217	0.2489	0.2703	0.2879	0.3026	0.3152	0.3261
	6260.6	99.4658	26.5045	13.8377	9.3793	7.2285	5.9920	5.1981	4.6486
60	0.1413	0.2009	0.2424	0.2740	0.2995	0.3206	0.3386	0.3542	0.3679
	6313.0	99.4825	26.3164	13.6522	9.2020	7.0567	5.8236	5.0316	4.4831
120	0.1460	0.2089	0.2532	0.2874	0.3151	0.3383	0.3582	0.3755	0.3908
	6339.4	99.4908	26.2211	13.5581	9.1118	6.9690	5.7373	4.9461	4.3978
∞	0.1507	0.2171	0.2644	0.3013	0.3314	0.3569	0.3789	0.3982	0.4154
	6365.9	99.4992	26.1252	13.4631	9.0204	6.8800	5.6495	4.8588	4.3105

Table A.4 continued

r	10	12	15	20	25	30	60	120	∞
					c				
1	1.7E-4	1.6E-4	1.6E-4	1.6E-4	1.6E-4	1.6E-4	1.6E-4	1.6E-4	1.6E-4
	10.0443	9.3302	8.6831	8.0960	7.7698	7.5625	7.0771	6.8509	6.6349
2	0.0101	0.0101	0.0101	0.0101	0.0101	0.0101	0.0101	0.0101	0.0101
	7.5594	6.9266	6.3589	5.8489	5.5680	5.3904	4.9774	4.7865	4.6052
3	0.0367	0.0370	0.0372	0.0375	0.0376	0.0377	0.0380	0.0381	0.0383
	6.5523	5.9525	5.4170	4.9382	4.6755	4.5097	4.1259	3.9491	3.7816
4	0.0687	0.0696	0.0704	0.0713	0.0719	0.0723	0.0732	0.0738	0.0743
	5.9943	5.4120	4.8932	4.4307	4.1774	4.0179	3.6490	3.4795	3.3192
5	0.0995	0.1011	0.1029	0.1047	0.1058	0.1066	0.1087	0.1097	0.1109
	5.6363	5.0643	4.5556	4.1027	3.8550	3.6990	3.3389	3.1735	3.0173
6	0.1270	0.1296	0.1323	0.1352	0.1371	0.1383	0.1417	0.1435	0.1453
	5.3858	4.8206	4.3183	3.8714	3.6272	3.4735	3.1187	2.9559	2.8020
7	0.1511	0.1546	0.1584	0.1625	0.1651	0.1669	0.1717	0.1743	0.1770
	5.2001	4.6395	4.1415	3.6987	3.4568	3.3045	2.9531	2.7918	2.6393
8	0.1720	0.1765	0.1813	0.1866	0.1900	0.1924	0.1987	0.2022	0.2058
	5.0567	4.4994	4.0045	3.5644	3.3239	3.1726	2.8233	2.6629	2.5113
9	0.1902	0.1956	0.2015	0.2080	0.2122	0.2151	0.2231	0.2274	0.2320
	4.9424	4.3875	3.8948	3.4567	3.2172	3.0665	2.7185	2.5586	2.4073
10	0.2062	0.2125	0.2194	0.2270	0.2320	0.2355	0.2450	0.2502	0.2558
	4.8491	4.2961	3.8049	3.3682	3.1294	2.9791	2.6318	2.4721	2.3209
12	0.2328	0.2407	0.2494	0.2592	0.2656	0.2702	0.2828	0.2899	0.2975
	4.7059	4.1553	3.6662	3.2311	2.9931	2.8431	2.4961	2.3363	2.1847
15	0.2628	0.2728	0.2839	0.2966	0.3050	0.3111	0.3282	0.3379	0.3486
	4.5581	4.0096	3.5222	3.0880	2.8502	2.7002	2.3523	2.1915	2.0385
20	0.2969	0.3095	0.3238	0.3404	0.3517	0.3599	0.3835	0.3973	0.4130
	4.4054	3.8584	3.3719	2.9377	2.6993	2.5487	2.1978	2.0346	1.8783
25	0.3195	0.3341	0.3509	0.3705	0.3840	0.3940	0.4231	0.4406	0.4610
	4.3111	3.7647	3.2782	2.8434	2.6041	2.4526	2.0984	1.9325	1.7726
30	0.3357	0.3517	0.3703	0.3924	0.4077	0.4191	0.4529	0.4738	0.4984
	4.2469	3.7008	3.2141	2.7785	2.5383	2.3860	2.0285	1.8600	1.6964
60	0.3800	0.4006	0.4251	0.4550	0.4766	0.4930	0.5446	0.5793	0.6247
	4.0819	3.5355	3.0471	2.6077	2.3637	2.2079	1.8363	1.6557	1.4730
120	0.4045	0.4280	0.4563	0.4915	0.5175	0.5376	0.6040	0.6523	0.7244
	3.9965	3.4494	2.9595	2.5168	2.2696	2.1108	1.7263	1.5330	1.3246
∞	0.4309	0.4577	0.4906	0.5324	0.5642	0.5895	0.6789	0.7550	1.0000
	3.9090	3.3608	2.8684	2.4212	2.1694	2.0062	1.6006	1.3805	1.0000

Table A.5 F-values for Two-Sided 99% $(1 - 2\alpha)$ or One-Sided 99.5% $(1 - \alpha)$
Intervals—$F_{.995:r,c}$ and $F_{.005:r,c}$

r	c 1	2	3	4	5	6	7	8	9
1	6.2E-5	5.0E-5	4.6E-5	4.4E-5	4.3E-5	4.3E-5	4.2E-5	4.2E-5	4.2E-5
	16210.7	198.5	55.5520	31.3328	22.7848	18.6350	16.2356	14.6882	13.6136
2	.005038	.005025	.005021	.005019	.005018	.005017	.005016	.005016	.005015
	19999.5	199.0	49.7993	26.2843	18.3138	14.5441	12.4040	11.0424	10.1067
3	0.0180	0.0201	0.0211	0.0216	0.0220	0.0223	0.0225	0.0227	0.0228
	21614.7	199.2	47.4672	24.2591	16.5298	12.9166	10.8824	9.5965	8.7171
4	0.0319	0.0380	0.0412	0.0432	0.0445	0.0455	0.0462	0.0468	0.0473
	22499.6	199.2	46.1946	23.1545	15.5561	12.0275	10.0505	8.8051	7.9559
5	0.0439	0.0546	0.0605	0.0643	0.0669	0.0689	0.0704	0.0716	0.0726
	23055.8	199.3	45.3916	22.4564	14.9396	11.4637	9.5221	8.3018	7.4712
6	0.0537	0.0688	0.0774	0.0831	0.0872	0.0903	0.0927	0.0946	0.0962
	23437.1	199.3	44.8385	21.9746	14.5133	11.0730	9.1553	7.9520	7.1339
7	0.0616	0.0806	0.0919	0.0995	0.1050	0.1092	0.1125	0.1152	0.1175
	23714.6	199.4	44.4341	21.6217	14.2004	10.7859	8.8854	7.6941	6.8849
8	0.0681	0.0906	0.1042	0.1136	0.1205	0.1258	0.1300	0.1334	0.1363
	23925.4	199.4	44.1256	21.3520	13.9610	10.5658	8.6781	7.4959	6.6933
9	0.0735	0.0989	0.1147	0.1257	0.1338	0.1402	0.1452	0.1494	0.1529
	24091.0	199.4	43.8824	21.1391	13.7716	10.3915	8.5138	7.3386	6.5411
10	0.0780	0.1061	0.1238	0.1362	0.1455	0.1528	0.1587	0.1635	0.1676
	24224.5	199.4	43.6858	20.9667	13.6182	10.2500	8.3803	7.2106	6.4172
12	0.0851	0.1175	0.1384	0.1533	0.1647	0.1737	0.1810	0.1871	0.1922
	24426.4	199.4	43.3874	20.7047	13.3845	10.0343	8.1764	7.0149	6.2274
15	0.0926	0.1299	0.1544	0.1723	0.1861	0.1972	0.2063	0.2139	0.2204
	24630.2	199.4	43.0847	20.4383	13.1463	9.8140	7.9678	6.8143	6.0325
20	0.1006	0.1431	0.1719	0.1933	0.2100	0.2236	0.2349	0.2445	0.2528
	24836.0	199.4	42.7775	20.1673	12.9035	9.5888	7.7540	6.6082	5.8318
25	0.1055	0.1516	0.1831	0.2068	0.2256	0.2410	0.2538	0.2648	0.2744
	24960.3	199.5	42.5910	20.0024	12.7554	9.4511	7.6230	6.4817	5.7084
30	0.1089	0.1574	0.1909	0.2163	0.2365	0.2532	0.2673	0.2793	0.2898
	25043.6	199.5	42.4658	19.8915	12.6556	9.3582	7.5345	6.3961	5.6248
60	0.1177	0.1726	0.2115	0.2416	0.2660	0.2864	0.3038	0.3190	0.3324
	25253.1	199.5	42.1494	19.6107	12.4024	9.1219	7.3088	6.1772	5.4104
120	0.1223	0.1805	0.2224	0.2551	0.2818	0.3044	0.3239	0.3410	0.3561
	25358.6	199.5	41.9895	19.4684	12.2737	9.0015	7.1933	6.0649	5.3001
∞	0.12690	0.1887	0.2337	0.2692	0.2985	0.3235	0.3452	0.3644	0.3815
	25464.9	199.5	41.8283	19.3247	12.1435	8.8793	7.0760	5.9506	5.1875

Table A.5 continued

r	10	12	15	20	c 25	30	60	120	∞
1	4.1E-5	4.1E-5	4.1E-5	4.0E-5	4.0E-5	4.0E-5	4.0E-5	3.9E-5	3.9E-5
	12.8265	11.7543	10.7981	9.9440	9.4753	9.1797	8.4946	8.1788	7.8794
2	.005015	.005015	.005014	.005014	.005014	.005013	.005013	.005013	.005013
	9.4270	8.5096	7.7008	6.9865	6.5982	6.3547	5.7950	5.5393	5.2983
3	0.0229	0.0230	0.0232	0.0234	0.0235	0.0235	0.0237	0.0238	0.0239
	8.0807	7.2258	6.4760	5.8177	5.4615	5.2388	4.7290	4.4972	4.2794
4	0.0477	0.0483	0.0489	0.0496	0.0500	0.0503	0.0510	0.0514	0.0517
	7.3428	6.5211	5.8029	5.1743	4.8351	4.6234	4.1399	3.9207	3.7151
5	0.0734	0.0747	0.0761	0.0775	0.0784	0.0790	0.0806	0.0815	0.0823
	6.8724	6.0711	5.3721	4.7616	4.4327	4.2276	3.7600	3.5482	3.3499
6	0.0976	0.0997	0.1019	0.1043	0.1058	0.1069	0.1096	0.1111	0.1126
	6.5446	5.7570	5.0708	4.4721	4.1500	3.9492	3.4918	3.2849	3.0913
7	0.1193	0.1223	0.1255	0.1290	0.1312	0.1327	0.1368	0.1390	0.1413
	6.3025	5.5245	4.8473	4.2569	3.9394	3.7416	3.2911	3.0874	2.8968
8	0.1387	0.1426	0.1468	0.1513	0.1543	0.1563	0.1619	0.1649	0.1681
	6.1159	5.3451	4.6744	4.0900	3.7758	3.5801	3.1344	2.9330	2.7444
9	0.1558	0.1606	0.1658	0.1715	0.1752	0.1778	0.1848	0.1887	0.1928
	5.9676	5.2021	4.5364	3.9564	3.6447	3.4505	3.0083	2.8083	2.6210
10	0.1710	0.1766	0.1828	0.1896	0.1941	0.1972	0.2058	0.2105	0.2156
	5.8467	5.0855	4.4235	3.8470	3.5370	3.3440	2.9042	2.7052	2.5188
12	0.1966	0.2038	0.2118	0.2208	0.2267	0.2309	0.2425	0.2491	0.2562
	5.6613	4.9062	4.2497	3.6779	3.3704	3.1787	2.7419	2.5439	2.3583
15	0.2261	0.2353	0.2457	0.2576	0.2655	0.2712	0.2873	0.2965	0.3067
	5.4707	4.7213	4.0698	3.5020	3.1963	3.0057	2.5705	2.3727	2.1868
20	0.2599	0.2719	0.2856	0.3014	0.3123	0.3202	0.3429	0.3564	0.3717
	5.2740	4.5299	3.8826	3.3178	3.0133	2.8230	2.3872	2.1881	1.9998
25	0.2827	0.2967	0.3129	0.3319	0.3451	0.3548	0.3833	0.4006	0.4208
	5.1528	4.4115	3.7662	3.2025	2.8981	2.7076	2.2698	2.0686	1.8771
30	0.2990	0.3146	0.3327	0.3542	0.3693	0.3805	0.4141	0.4348	0.4596
	5.0706	4.3309	3.6867	3.1234	2.8187	2.6278	2.1874	1.9840	1.7891
60	0.3443	0.3647	0.3890	0.4189	0.4406	0.4572	0.5096	0.5452	0.5922
	4.8592	4.1229	3.4803	2.9159	2.6088	2.4151	1.9622	1.7469	1.5325
120	0.3697	0.3931	0.4215	0.4570	0.4834	0.5040	0.5725	0.6229	0.6988
	4.7501	4.0149	3.3722	2.8058	2.4961	2.2998	1.8341	1.6055	1.3637
∞	0.3970	0.4240	0.4573	0.5000	0.5327	0.5590	0.6525	0.7333	1.0000
	4.6385	3.9039	3.2602	2.6904	2.3765	2.1760	1.6885	1.4311	1.0000

APPENDIX B

Wald's Confidence Interval on a Ratio of Two Variance Components

In a series of three papers, Wald (1940, 1941, 1947) proposed a procedure for constructing exact confidence intervals on certain ratios of variance components in unbalanced experimental designs. Seely and El-Bassiouni (1983) generalized this approach and derived conditions under which the method can be applied. Harville and Fenech (1985) considered computational aspects of the problem and also derived approximate intervals that are simple to compute. Using the notation of Harville and Fenech, consider the mixed model with one set of random effects (in addition to the set of residual effects)

$$\underline{Y} = X\underline{\beta} + Z\underline{S} + \underline{E} \tag{B.1}$$

where $\underline{Y}$ is an $N \times 1$ vector of observations, $\underline{\beta}$ is a $p \times 1$ vector of unknown parameters, $\underline{S}$ and $\underline{E}$ are $m \times 1$ and $N \times 1$ independent multivariate normal random vectors with mean vector $\underline{0}$ and variance-covariance matrices $\sigma_S^2 I_m$ and $\sigma_E^2 I_N$, respectively, and I_m and I_N are identity matrices of order m and N, respectively. The matrices X and Z are of dimensions $N \times p$ and $N \times m$, respectively, and represent design matrices for the fixed and random effects. Finally, define $p^* =$

rank(X), $r = $ rank(X, Z) $- p^*$, and $f = N - r - p^*$. The problem of interest is to construct an exact confidence interval on the ratio $\lambda = \sigma_S^2/\sigma_E^2$. In order for this to be possible, we must have $r > 0$ and $f > 0$.

To illustrate the notation, consider the one-fold nested design in (4.1.1). In terms of the model in (B.1), $p = 1$, $m = I$, $\underline{Y}$ corresponds to an $N \times 1$ vector of the Y_{ij}, X is an $N \times 1$ vector of 1s, $\beta = \mu$, $\underline{S}$ is an $I \times 1$ vector of the A_i random variables, and $\underline{E}$ is an $N \times 1$ vector of the E_{ij} random variables. The design matrix Z is

$$
Z = \begin{bmatrix} 1 & 0 & \cdots & 0 \\ 0 & 1 & \cdots & 0 \\ \vdots & \vdots & \ddots & \vdots \\ 0 & 0 & \cdots & 1 \end{bmatrix} \begin{matrix} \}J_1 \text{ rows} \\ \}J_2 \text{ rows} \\ \vdots \\ \}J_I \text{ rows} \end{matrix}
$$

For this design, $r = I - 1$, $p^* = 1$, and $f = N - I$.

The exact two-sided $1 - 2\alpha$ confidence interval on λ proposed by Wald is $[L^*; U^*]$ where L^* is given by the root of the equation

$$
G(\lambda) = F_{\alpha:r,f} \tag{B.2}
$$

and U^* is given by the root of the equation

$$
G(\lambda) = F_{1-\alpha:r,f} \tag{B.3}
$$

where

$$
G(\lambda) = \frac{f\underline{t}'(I_r + \lambda D)^{-1}\underline{t}'}{r(SSE)} = \frac{f\sum_i^r t_i^2/(1 + \lambda\Delta_i)}{r(SSE)} \tag{B.4}
$$

$$
D = \text{diag}(\Delta_1, \Delta_2, \ldots, \Delta_r) \qquad \text{with } 0 < \Delta_1 \leq \ldots \leq \Delta_r,
$$

$$
\Delta_i = \text{nonzero eigenvalue of } C; \qquad i = 1, \ldots, r
$$

$$
C = Z'(I_N - X(X'X)^G X')Z
$$

$(X'X)^G$ is a generalized inverse of $X'X$

V is an $m \times r$ matrix with columns that are orthonormal characteristic vectors of C

$$\underline{t}' = D^{-1/2}V'Z'(I_N - X(X'X)^G X')\underline{Y} \quad \text{and}$$

$$SSE = \underline{Y}'\underline{Y} - \underline{Y}'(X(X'X)^G X')\underline{Y} - \underline{t}'\underline{t}$$

The function $G(\lambda)$ is a convex and monotonically nonincreasing function of λ for $0 \le \lambda < \infty$ with $G(0) = f\sum_i^r t_i^2/(rSSE)$. L^* is defined to zero if $G(0) < F_{\alpha:r,f}$. If $G(0) < F_{1-\alpha:r,f}$, L^* and U^* are both defined to be zero.

In order to compute the exact interval, one must solve the two non-linear equations in (B.2) and (B.3). Burdick, Maqsood, and Graybill (1986) illustrate how the bisection method can be used to easily solve these equations. The bisection method allows the investigator to determine the precision of the estimate after any iteration and is easy to program. To demonstrate this method, we define the bounds

$$L_1 = \frac{f\sum_i^r t_i^2/\Delta_i}{r(SSE)F_{\alpha:r,f}} - \frac{1}{\Delta_1} \tag{B.5}$$

$$L_r = \frac{f\sum_i^r t_i^2/\Delta_i}{r(SSE)F_{\alpha:r,f}} - \frac{1}{\Delta_r}$$

$$U_1 = \frac{f\sum_i^r t_i^2/\Delta_i}{r(SSE)F_{1-\alpha:r,f}} - \frac{1}{\Delta_1} \quad \text{and}$$

$$U_r = \frac{f\sum_i^r t_i^2/\Delta_i}{r(SSE)F_{1-\alpha:r,f}} - \frac{1}{\Delta_r}$$

These bounds are obtained by rewriting (B.4) as

$$G(\lambda) = \frac{f\sum_i^r (\Delta_i^{-1} + \lambda)^{-1}t_i^2/\Delta_i}{r(SSE)} \tag{B.6}$$

Replacement of the term $(\Delta_i^{-1} + \lambda)^{-1}$ with $(\Delta_1^{-1} + \lambda)^{-1}$ provides a function $G_1(\lambda)$ such that $G_1(\lambda) \le G(\lambda)$. Similarly, if $(\Delta_i^{-1} + \lambda)^{-1}$ is replaced with $(\Delta_r^{-1} + \lambda)^{-1}$, the resulting function $G_r(\lambda)$ satisfies $G(\lambda) \le G_r(\lambda)$. By setting $G_1(\lambda)$ and $G_r(\lambda)$ equal to appropriate F-values, the bounds in (B.5) are obtained. Additionally, since $G_1(\lambda) \le G(\lambda) \le$

$G_r(\lambda)$, then $L_1 \le L^* \le L_r$ and $U_1 \le U^* \le U_r$. The bisection method uses these results to compute L^* and U^*.

An algorithm for computing the upper bound on λ based on the bisection method is as follows:

(i)   If $G(0) < F_{1-\alpha:r,f}$, then define $U^* = 0$. If $G(0) > F_{1-\alpha:r,f}$, then go to (ii).

(ii)   Let $a = \text{Max}(0, U_1)$, $b = U_r$, and $n = 1$. Go to (iii).

(iii)   Compute $c_n = (a + b)/2$. It must be true that $|U^* - c_n| \le (1/\Delta_1 - 1/\Delta_r)/2^n$. Thus, if $(1/\Delta_1 - 1/\Delta_r)/2^n$ is sufficiently small, stop and use c_n as the upper bound. Otherwise, proceed to (iv).

(iv)   Compute $g(c_n) = G(c_n) - F_{1-\alpha:r,f}$ where $G(c_n)$ is defined in (B.4) by replacing λ with c_n. If $g(c_n) > 0$, set $a = c_n$, $n = n + 1$, and return to (iii). If $g(c_n) < 0$, set $b = c_n$, $n = n + 1$, and return to (iii). If $g(c_n) = 0$, an exact solution has been obtained.

The same procedure is followed to find a lower bound, d_n, on λ. If $G(0) < F_{\alpha:r,f}$, then $L^* = 0$. If $G(0) > F_{\alpha:r,f}$, set $a = \text{Max}(0, L_1)$, $b = L_r$, $c_n = d_n$, $g(d_n) = G(d_n) - F_{\alpha:r,f}$, and begin at step (iii).

We illustrate the bisection method algorithm by computing a 90% confidence interval on σ_A^2/σ_E^2 for the bull data in Table 4.3.1. The following code is written in SAS©/IML.

```
DATA BULLS;
INPUT BULL Y;
CARDS;
1 46
1 31
....
DATA GOES HERE
....
6 29
6 60
PROC IML;
START;
PREC=.0001; *THE DIFFERENCE BETWEEN CN AND L* AND DN AND U*
IS LESS THAN PREC;
ALPHA=.05;BETA=1-ALPHA; *A TWO-SIDED INTERVAL HAS CONFIDENCE
COEFFICIENT OF 1-2*ALPHA;
*####################;
```

```
*THE FOLLOWING STATEMENTS WILL VARY DEPENDING ON THE DESIGN
THAT IS EMPLOYED;
*THE STATEMENTS USED HERE ARE FOR A ONE-FOLD NESTED DESIGN;
USE BULLS;READ ALL INTO MAT;
Y=MAT(|,2|);
*Z IS THE DESIGN MATRIX OF THE RANDOM EFFECTS;
Z=DESIGN(MAT(|,1|));
X IS THE DESIGN MATRIX OF THE FIXED EFFECTS;
X=J(NROW(Z),1,1);
*###########################;
N=NROW(Z);
*DEFINE THE MATRICES C AND Q;
C=Z'*(I(N)-X*GINV(X'*X)*X')*Z;
Q=Z'*(I(N)-X*GINV(X'*X)*X')*Y;
CALL EIGEN(DE,VE,C);
DR=DE>.0001; *THIS CORRECTS FOR ROUNDOFF ON EIGENVALUES;
XE=EIGVAL(X'*X);
XC=XE>.0001; *THIS CORRECTS FOR ROUNDOFF ON EIGENVALUES;
*COMPUTE RANKS AND DEGREES OF FREEDOM;
R=SUM(DR);
PSTAR=SUM(XC);
F=N-R-PSTAR;
*COMPUTE F VALUES FOR (B.2) AND (B.3);
FU=FINV(BETA,R,F);
FL=FINV(ALPHA,R,F);
D=DIAG(DE(|1:R,|));
DELTA1=MIN(DE(|1:R,|));
DELTAR=MAX(DE(|1:R,|));
V=VE(|,1:R|);
T=INV(D##.5)*V'*Q;
SSE=Y'*Y-Y'*X*GINV(X'*X)*X'*Y-T'*T;
K=F/(R*SSE);
L1=K*T'*INV(D)*T/FU-1/DELTA1;
LR=L1+1/DELTA1-1/DELTAR;
UR=K*T'*INV(D)*T/FL-1/DELTAR;
U1=UR+1/DELTAR-1/DELTA1;
*COMPUTE EXACT UPPER BOUND;
AC=MAX(0,U1);BC=UR;
PRECNT=INT(LOG((1/DELTA1-1/DELTAR)/PREC)/LOG(2)+1);
IF (T'*T*F)/(SSE*R)<FL THEN CN=0;
ELSE DO CNT=1 TO PRECNT;
CN=(AC+BC)/2;
IF K*T'*INV(I(R)+CN*D)*T-FL>0 THEN AC=CN;
```

```
ELSE BC=CN;
END;USTR=CN;PRINT USTR;
*COMPUTE EXACT LOWER BOUND;
AC=MAX(0,L1);BC=LR;
IF(T'*T*F)/(SSE*R)<FU THEN DN=0;
ELSE DO CNT=1 TO PRECNT;
DN=(AC+BC)/2;
IF K*T'*INV(I(R)+DN*D)*T-FU>0 THEN AC=DN;
ELSE BC=DN;
END;
LSTR=DN;PRINT LSTR;
FINISH;RUN;
```

For this data set $N = 35, r = 5, p^* = 1, f = 29, F_{.05:5,29} = 2.5454$, and $F_{.95:5,29} = .2222$. The computed values are $\sum_i^r t_i^2/\Delta_i = 659.3$, $SSE = 7200.3, \Delta_1 = 2.2525, \Delta_5 = 8.3441, U_1 = 1.9462, U_5 = 2.2701, L_1 = -.2353, L_5 = .0888, c_{12} = .0092, d_{12} = 2.1584$, with $n = 12$ iterations. The 12 iterations guarantee that both $| U^* - c_n |$ and $| L^* - d_n |$ are less than .0001.

The inequalities $L_1 \leq L^* \leq L_r$ and $U_1 \leq U^* \leq U_r$ imply that a conservative two-sided $1 - 2\alpha$ confidence interval on λ is

$$[L_1 ; U_r] \tag{B.7}$$

In situations where the Δ_i can be bounded by quantities that are easily obtained from the experimental design, (B.7) can be modified to provide a conservative interval that requires no computer computations. For example, LaMotte (1976) showed that in the one-fold nested design $Min(J_i) \leq \Delta_1 \leq \cdots \leq \Delta_r \leq Max(J_i)$. Thus, replacement of Δ_1 with $Min(J_i)$ and Δ_r with $Max(J_i)$ provides a conservative interval on λ that doesn't require the computation of eigenvalues. Kala, Molińska, and Moliński (1990) provide tighter conservative intervals than (B.7). However, these intervals require calculation of eigenvalues. If one must perform this computation, the exact interval might as well be obtained.

Harville and Fenech (1985) show that model (B.1) can be generalized to include more than one set of random effects. Very simply, if all random effects except those contained in $\underline{S}$ are placed in $\underline{\beta}$ and treated as fixed effects, the same procedures can be applied. Of course, it must be assumed that the rank conditions $r > 0$ and $f > 0$ are satisfied. Examples of models with more than one set of random effects are provided in Sections 5.4.3 and 6.5.

Power comparisons of (B.2) and (B.3) with other test statistics have been conducted by El-Bassiouni and Seely (1988), LaMotte, McWhorter, and Prasad (1988), Westfall (1988, 1989), and Lin and Harville (1991).

References

Aastveit, A.H. (1990). Use of bootstrapping for estimation of standard deviation and confidence intervals of genetic variance- and covariance-components. *Biomet. J.*, *32*, 515–527. R1

Anderson, T. W. (1985). Components of variance in MANOVA. *Multi. Anal. VI*, P. R. Krishnaiah (Ed), North-Holland Publishing, 1–8. R1.

Arteaga, C., Jeyaratnam, S., and Graybill, F.A. (1982). Confidence intervals for proportions of total variance in the two-way cross component of variance model. *Comm. Statist. - Theor. Meth.*, *11*, 1643–1658. R6

Arvesen, J.N. (1969). Jackknifing U-statistics. *Ann. Math. Stat.*, *40*, 2076–2100. R1

Arvesen, J.N. and Layard, M.W.J. (1975). Asymptotically robust tests in unbalanced variance component models. *Ann. Stat.*, *3*, 1122–1134. R1, R4

Arvesen, J.N. and Schmitz, T.H. (1970). Robust procedures for variance component problems using the jackknife. *Biometrics*, *26*, 677–686. R1

Balakrishnan, N. and Ma, C.W. (1990). A comparative study of various tests for the equality of two population variances. *J. Stat. Comp. and Sim.*, *35*, 41–89. R2

KEY: R = Chapters in which paper is referenced. * = Paper forthcoming or in review.

Banerjee, S.K. (1960). Approximate confidence interval for linear functions of means of k populations when the population variances are not equal. *Sankhyā*, *22*, 357–358. R7

Bartlett, M.S. (1953). Approximate confidence intervals II. More than one unknown parameter. *Biometrika*, *40*, 306–317. R3

Bhargava, R.P. (1946). Test of significance for intraclass correlation when family sizes are not equal. *Sankhyā*, *7*, 435–438. R4

Birch, N.J., and Burdick, R.K. (1989). Confidence intervals on the ratio of expected mean squares $(\theta_1 + \theta_2 + \theta_3)/\theta_4$. *Stat. and Prob. Lett.*, *7*, 335–342. R3,R6

Birch, N.J., Burdick, R.K., and Ting, N. (1990). Confidence intervals and bounds for a ratio of summed expected mean squares. *Technometrics*, *32*, 437–444. R3,R4,R6

Bliss, C.I. (1967). *Statistics in Biology, Volume 1*, McGraw-Hill Book Company, New York. R5

Boardman, T.J. (1974). Confidence intervals for variance components-A comparative Monte Carlo study. *Biometrics*, *30*, 251–262. R3

Box, G.E.P. and Tiao, G.C. (1973). *Bayesian Inference in Statistical Analysis*, Addison-Wesley, Reading, Mass. R1

Broemeling, L.D. (1969a). Confidence intervals for measures of heritability. *Biometrics*, *25*, 424–427. R3,R5,R6

Broemeling, L.D. (1969b). Confidence regions for variance ratios of random models. *J. Amer. Stat. Assoc.*, *64*, 660–664. R3,R5,R6

Broemeling, L.D. (1978). Simultaneous inferences for variance ratios of some mixed linear models. *Comm. Stat. - Theor. Meth.*, *7*, 297–305. R3

Broemeling, L.D. (1985). *Bayesian Analysis of Linear Models*, Marcel Dekker, New York. R1

Broemeling, L.D. and Bee, D.E. (1976). Simultaneous confidence intervals for parameters of a balanced incomplete block. *J. Amer. Stat. Assoc.*, *71*, 425–428. R6

Bross, I. (1950). Fiducial intervals for variance components. *Biometrics*, *6*, 136–144. R1

Brownlee, K. A. (1953). *Industrial Experimentation*, Chemical Publishing Co., Inc., New York. R5,R7

Bulmer, M.G. (1957). Approximate confidence limits for components of variance. *Biometrika, 44*, 159–167. R1, R3

Burdick, R.K., Birch, N.J., and Graybill, F.A. (1986). Confidence intervals on measures of variability in an unbalanced two-fold nested design with equal subsampling. *J. Statist. Comput. Simul., 25*, 259–272. R5

Burdick, R.K. and Eickman, J. (1986). Confidence intervals on the among group variance component in the unbalanced one-fold nested design. *J. Stat. Comput. Simul., 26*, 205–219. R4

Burdick, R.K. and Graybill, F.A. (1984). Confidence intervals on linear combinations of variance components in the unbalanced one-way classification. *Technometrics, 26*, 131–136. R4

Burdick, R.K. and Graybill, F.A. (1985). Confidence intervals on the total variance in an unbalanced two-fold nested classification with equal subsampling. *Comm. Stat. - Theor. Meth., 14*, 761–774. R5

Burdick, R.K. and Graybill, F.A. (1988). The present status of confidence interval estimation on variance components in balanced and unbalanced random models. *Comm. Stat. - Theor. Meth., 17*, 1165–1195. R1

Burdick, R.K., Maqsood, F., and Graybill, F.A. (1986). Confidence intervals on the intraclass correlation in the unbalanced one-way classification. *Comm. Stat. - Theor. Meth., 15*, 3353–3378. R4,RB

Burdick, R.K. and Sielken Jr., R.L. (1978). Exact confidence intervals for linear combinations of variance components in nested classifications. *J. Amer. Stat. Assoc., 73*, 632–635. R3,R7

Burk, F., Dion, L., Fridshal, D., Langford, E., O'Cinneide, C. and Parsons, T. (1984). On a conjecture relating χ^2 and F quantiles. *Comm. Stat.-Theor. Meth., 13*, 661–670. R3

Calvin, J.A., Jeyaratnam, S., and Graybill, F.A. (1986). Approximate confidence intervals for the three-factor mixed model. *Comm. Stat. - Simula. Computa., 15*, 893–903. R7

Cochran, W.G. (1951). Testing a linear relation among variances. *Biometrics, 7*, 17–32. R3,R6

Crump, S.L. (1946). The estimation of variance components in analysis of variance. *Biometrics Bulletin, 2*, 7–11. R3

Crump, S.L. (1951). The present status of variance component analysis. *Biometrics, 7*, 1–16. R1

Cummings, W.B., and Gaylor, D.W. (1974). Variance component testing in unbalanced nested designs. *J. Amer. Stat. Assoc.*, *69*, 765–771. R5

Daniels, H.E. (1939). The estimation of components of variance. *J. Roy. Stat. Soc.*, *Suppl.*, *6*, 186–197. R1

Davenport, James M. (1975). Two methods of estimating the degrees of freedom of an approximate F. *Biometrika*, *62*, 682–684. R3,R6

Davenport, James M. and Webster, J.T. (1973). A comparison of some approximate F-tests. *Technometrics*, *15*, 779–789. R6

Donner, A. (1986). A review of inference procedures for the intraclass correlation coefficient in the one-way random effects model. *Int. Stat. Rev.*, *54*, 67–82. R4

Donner, A. and Eliasziw, M. (1987). Sample size requirements for reliability studies. *Stat. in Med.*, *6*, 441–448. R4

Donner, A. and Koval, J.J. (1989) The effect of imbalance on significance-testing in one-way model II analysis of variance. *Comm. Stat.-Theor. Meth.*, *18(4)*, 1239–1250. R4

Donner, A. and Wells, G. (1986). A comparison of confidence interval methods for the intraclass correlation coefficient. *Biometrics*, *42*, 401–412. R4

Donner, A., Wells, G., and Eliasziw, M. (1989). On two approximations to the F-distribution: Application to testing for intraclass correlation in family studies. *Canad. J. Stat.*, *17*, 209–215. R4

Donoghue, J.R. and Collins, L.M. (1990). A note on the unbiased estimation of the intraclass correlation. *Psychometrika*, *55*, 159–164. R4

El-Bassiouni, M.Y. and Seely, J.F. (1988). On the power of Wald's variance component test in the unbalanced random one-way model. *Optimal Design and Analysis of Experiments*, North-Holland Publishing, 157–165. RB

Fisher, R.A. (1924). On a distribution yielding the error functions of several well known statistics. *Proc. Internat. Math. Congr.*, Toronto, 805–813. R1

Fleiss, J.L. and Shrout, P.E. (1978). Approximate interval estimation for a certain intraclass correlation coefficient. *Psychometrika 43*, 259–262. R6

Gallo, J. and Khuri, A.I. (1990). Exact tests for the random and fixed effects in an unbalanced mixed two-way cross-classification model. *Biometrics*, *46*, 1087–1095. R7

Gianola, D. (1980). Confidence intervals for ratios of linear functions of mixed models with reference to animal breeding data. *J. Anim. Sci.*, *50*, 1051–1056. R7

Graybill, F.A. (1961). *An Introduction to Linear Statistical Models*, Mc-Graw-Hill, New York. R5, R6

Graybill, F.A. (1976). *Theory and Application of the Linear Model*, Duxbury, North Scituate, Massachusetts. R2, R6

Graybill, F. A., and Hultquist, R. A. (1961). Theorems concerning Eisenhart's Model II. *Ann. Math. Statist.*, *32*, 261–269. R1,R2

Graybill, F.A. and Robertson, W.H. (1957). Calculating confidence intervals for genetic heritability. *Poultry Sci.*, *36*, 261–265. R5

Graybill, F.A. and Wang, C.M. (1979). Confidence intervals for proportions of variability in two-factor nested variance component models. *J. Amer. Stat. Assoc.*, *74*, 368–374. R5

Graybill, F.A. and Wang, C.M. (1980). Confidence intervals on nonnegative linear combinations of variances. *J. Amer. Stat. Assoc.*, *75*, 869–873. R3,R4

Green, J.R. (1954). A confidence interval for variance components. *Ann. Math. Stat.*, *25*, 671–686. R1

Groggel, D.J., Wackerly, D.D., and Rao, P.V. (1988). Nonparametric estimation in one-way random effects models. *Comm. Stat. Simula.*, *17*, 887–903. R4

Guenther, W.C. (1972). On the use of the incomplete gamma table to obtain unbiased tests and unbiased confidence intervals for the variance of a normal distribution. *Amer. Stat.*, *26*, 31–34. R2

Guenther, W.C. (1977). Calculation of factors for tests and confidence intervals concerning the ratio of two normal variances. *Amer. Stat.*, *31*, 175–177. R2

Hartley, H.O. and Rao, J.N.K. (1967). Maximum likelihood estimation for the mixed analysis of variance model. *Biometrika*, *54*, 93–108. R3

Hartung, J. and Voet, B. (1987). An asymptotic χ^2-test for variance components. *Contributions to Stochastics*, W. Sendler (Ed.), Physica-Verlag, Heidelberg, 153–163. R5

Harville, D.A. (1976). Confidence intervals and sets for linear combinations of fixed and random effects. *Biometrics*, *32*, 403–407. R7

Harville, D.A. and Fenech, A.P. (1985). Confidence intervals for a variance ratio, or for heritability, in an unbalanced mixed linear model. *Biometrics*, *41*, 137–152. RB

Healy, W.C., Jr. (1961). Limits for a variance component with an exact confidence coefficient. *Ann. Math. Stat.*, *32*, 466–476. R3

*Hernandez, R.P., Burdick, R.K., and Birch, N.J. (1992). Confidence intervals and tests of hypotheses on variance components in an unbalanced two-fold nested design. Forthcoming in *Biom. J.* R5

*Hernandez, R.P., and Burdick, R.K. (1992). Confidence intervals on the total variance in an unbalanced two-fold nested design. Paper in review. R5

Hernandez, R.P. (1991). Confidence intervals on linear combinations of variance components in unbalanced designs. Unpublished dissertation, Arizona State University, Tempe, Arizona. R6

Hocking, R.R. (1973). A discussion of the two-way mixed model. *Amer. Stat.*, *27*, 148–152. R7

Howe, W.G. (1974). Approximate confidence limits on the mean of $X + Y$ where X and Y are two tabled independent random variables. *J. Amer. Stat. Assoc.*, *69*, 789–794. R3,R6

Howe, R.B. and Myers, R.H. (1970). An alternative to Satterthwaite's test involving positive linear combinations of variance components. *J. Amer. Stat. Assoc.*, *65*, 404–412. R3

Huitson, A. (1955). A method of assigning confidence limits to linear combinations of variances. *Biometrika*, *42*, 471–479. R1,R3

Hussein, M. and Milliken, G.A. (1978a). An unbalanced two-way model with random effects having unequal variances. *Biom. J.*, *20*, 203–213. R6

Hussein, M. and Milliken, G.A. (1978b). An unbalanced nested model with random effects having unequal variances. *Biom. J.*, *20*, 329–338. R5

Imhof, J.P. (1960). A mixed model for the complete three-way layout with two random-effects factors. *Ann. Math. Stat.*, *31*, 906–928. R7

Jeske, D.R. and Harville, D.A. (1988). Prediction-interval procedures and (fixed effects) confidence interval procedures for mixed linear models. *Comm. Stat.-Theor. Meth.*, *17*, 1053–1087. R7

Jeyaratnam, S. and Graybill, F.A. (1980). Confidence intervals on variance components in three-factor cross-classification models. *Technometrics*, *22*, 375–380. R3,R6

Jeyaratnam, S. and Othman, A.R. (1985). Test of hypothesis in one-way random effects model with unequal error variances. *J. Stat. Comput. Simul.*, *21*, 51–57. R4

Jeyaratnam, S. and Panchapakesan, S. (1988). Prediction intervals in balanced one-factor random model. *Probability and Statistics: Essays in Honor of Franklin A. Graybill*, J.N. Srivastava (Ed.), North-Holland Publishing Co., 161–170. R7

John, S. (1973). Critical values for inference about normal dispersion. *Austral. J. of Stat.*, *15*, 71–79. R2

Johnson, N. L., and Leone, F. C. (1977). *Statistics and Experimental Design in Engineering and the Physical Sciences, Second edition*, John Wiley & Sons, New York. R6,R7

Kala, R., Molińska, A., and Moliński, K. (1990). Approximate confidence intervals for a ratio of two variance components. *Comm. Stat.-Theor. Meth.*, *19*, 1889–1898. RB

Kazempour, M.K. and Graybill, F.A. (1989). Confidence bounds for proportion of σ_E^2 in unbalanced two-way crossed models. *Statistical Data Analysis*, Y. Dodge (Ed), North-Holland Publishing Co., 389–396. R6

*Kazempour, M.K. and Graybill, F.A. (1991a). Approximate confidence bounds for ratios of variance components in two-way unbalanced crossed models with interactions. Forthcoming in *Comm. Stat.-Simul. Comp.* R6

*Kazempour, M.K. and Graybill, F.A. (1991b). Confidence bounds for variance components in unbalanced two-way crossed models. Paper in review. R6

*Kazempour, M.K. and Graybill, F.A. (1992). Confidence intervals on individual variances in two-way models. Forthcoming in *Comp. Stat. and Data Analysis*. R6

Khuri, A.I. (1981). Simultaneous confidence intervals for functions of variance components in random models. *J. Amer. Stat. Assoc.*, *76*, 878–885. R2,R3,R5

Khuri, A.I. (1984). Interval estimation of fixed effects and of functions of variance components in balanced mixed models. *Sankhyā*, Series B, 46, 10–28. R4

Khuri, A.I. (1987). An exact test for the nesting effect's variance component in an unbalanced random two-fold nested model. *Stat. and Prob. Lett.*, *5*, 305–311. R5

Khuri, A.I. (1990). Exact tests for random models with unequal cell frequencies in the last stage. *J. of Stat. Plan. and Infer.*, *24*, 177–193. R5.

Khuri, A.I. and Littell, R.C. (1987). Exact tests for the main effects variance components in an unbalanced random two-way model. *Biometrics*, *43*, 545–560. R6,R7

Khuri, A.I. and Sahai, H. (1985). Variance components analysis: A selective literature survey. *Int. Stat. Rev.*, *53*, 279–300. R1,R2

Kirk, R. E. (1982). *Experimental design*, Second edition, Brooks/Cole Publishing Co., Belmont, CA. R2

Klein, S.W. (1990). Effect of sample size on width of confidence intervals for variance and variance ratios in normal populations. *Comp. Stat. Quart.*, *3*, 171–179. R2

Kolodziejczyk, S. (1935). On an important class of statistical hypotheses. *Biometrika*, *27*, 161–190. R1

Konishi, S. and Khatri, C.G. (1990). Inferences on interclass and intraclass correlations in multivariate familial data. *Ann. Inst. Stat. Math.*, *42*, 561–580. R4

LaMotte, L.R. (1976). Invariant quadratic estimators in the random, one-way ANOVA model. *Biometrics*, *32*, 793–804. RB

LaMotte, L.R., McWhorter, A., and Prasad, R.A. (1988). Confidence intervals and tests on the variance ratio in random models with two variance components. *Comm. Stat. - Theor. Meth.*, *17*, 1135–1164. RB

Lehmann, E.L. (1986). *Testing Statistical Hypotheses*, John Wiley & Sons, New York. R4,R5

Leiva, R.A. and Graybill, F.A. (1986). Confidence intervals for variance components in the balanced two-way model with interaction. *Comm. Stat.-Simul. Comp.*, *15*, 301–322, R6

Levy, K.J. and Narula, S.C. (1974). Shortest confidence intervals for the ratio of two normal variances. *Canad. J. Stat.*, *2*, 83–87. R2

Limam, M.M.T. and Thomas, D.R. (1988). Simultaneous tolerance intervals in the random one-way model with covariates. *Comm. Stat.-Simul.*, *17*, 1007–1019. R1

Lin, P.K.H., Richards, D.O., Long, D.R., Myers, M.D., and Taylor, J.A. (1983). Tables for computing shortest confidence intervals involving the F-distribution. *Comm. Stat.-Simul.*, *17*, 711–725. R2

Lin, T.-H. and Harville, D.A. (1991). Some alternatives to Wald's confidence interval and test. *J. Amer. Stat. Assoc.*, *86*, 179–187. RB

Lindley, D.V., East, D.A., and Hamilton, P.A. (1960). Tables for making inferences about the variance of a normal distribution. *Biometrika*, *47*, 433–437. R2

Lu, T.-F.C. (1985). Confidence intervals on sums, differences and ratios of variance components. Unpublished dissertation, Colorado State University, Fort Collins, Colorado. R3

Lu, T.-F.C., Graybill, F.A., and Burdick, R.K. (1987). Confidence intervals on the ratio of expected mean squares $(\theta_1 + d\theta_2)/\theta_3$. *Biometrics*, *43*, 535–543. R3,R5,R6

Lu, T.-F.C., Graybill, F.A., and Burdick, R.K. (1988). Confidence intervals on a difference of expected mean squares. *J. Stat. Plan. and Infer.*, *18*, 35–43. R3

Lu,T.-F.C., Graybill, F.A., and Burdick, R.K. (1989). Confidence intervals on the ratio of expected mean squares $(\theta_1 - d\theta_2)/\theta_3$. *J. Stat. Plan. and Infer.*, *21*, 179–190. R3

Marcuse, S. (1949). Optimum allocation and variance components in nested sampling with an application to chemical analysis. *Biometrics*, *5*, 189–206. R7

Mathew, T. and Sinha, B.K. (1988a). Optimum tests for fixed effects and variance components in balanced models. *J. Amer. Stat. Assoc.*, *83*, 133–135. R7

Mathew, T. and Sinha, B.K. (1988b). Optimum tests in unbalanced two-way models without interaction. *Ann. of Stat.*, *16*, 1727–1740. R6

Mee, R.W. (1984). β-expectation and β-content tolerance limits for balanced one-way ANOVA random model. *Technometrics*, *26*, 251–254. R1

Mee, R.W. and Owen, D.B. (1983). Improved factors for one-sided tolerance limits for balanced one-way ANOVA random model. *J. Amer. Stat. Assoc.*, *78*, 901–905. R1

Mian, I.U.H., Shoukri, M.M., and Tracy, D.S. (1989). A comparison of significance testing procedures for the intraclass correlation from family data. *Comm. Stat.-Simul. Comp.*, *18*, 613–631. R4

Miller, R.G., Jr. (1981). *Simultaneous Statistical Inference*, Second Edition, Springer-Verlag, New York. R2

Milliken, G.A., and Johnson, D.E. (1984). *Analysis of Messy Data*, Volume 1, Lifetime Learning Publications, Belmont, CA. R2,R6,R7

Molińska, A. and Moliński, K. (1989). Interval estimation of parameters in an unbalanced mixed two-fold nested classification: The last stage uniformity case. *Comm. Stat.-Theor. Meth.*, *18*, 4121–4135. R7

Montgomery, D.C. (1984). *Design and Analysis of Experiments*, Second Edition, John Wiley & Sons, New York. R7

Moriguti, S. (1954). Confidence limits for a variance component. *Rep. Stat. Appl. Res.*, *JUSE*, *3*, 7–19. R1,R3

Mostafa, M.G. (1967). Note on testing hypotheses in an unbalanced random effects model. *Biometrika*, *54*, 659–662. R4

Mostafa, S.M. and Ahmad, R. (1986). Confidence intervals for variance components in balanced random models when the errors are independent and related through an autoregressive series. *Statistica*, *46*, 363–377. R3

Myers, R.H. and Howe, R.B. (1971). On alternative approximate F tests for hypotheses involving variance components. *Biometrika*, *58*, 393–396. R3,R6

Nagata, Y. (1989). Improvements of interval estimations for the variance and the ratio of two variances. *J. of the Jap. Stat. Soc.*, *19*, 151–161. R2

Naik, U.D. (1974). On tests of main effects and interactions in higher-way layouts in the analysis of variance random effects model. *Technometrics*, *16*, 17–25. R6,R7

Olkin, I. and Pratt, J.W. (1958). Unbiased estimation of certain correlation coefficients. *Ann. Math. Stat.*, *29*, 201–211. R4

Pachares, J. (1961). Tables for unbiased tests on the variance of a normal population. *Ann. Math. Stat.*, *32*, 84–87. R2

Prasad, N.G.N. and Rao, J.N.K. (1988). Robust tests and confidence intervals for error variance in a regression model and for functions of variance components in an unbalanced random one-way model. *Comm. Stat.-Theor. Meth.*, *17*, 1111–1133. R1,R4

Ramachandran, K.V. (1956). Contributions to simultaneous confidence interval estimation. *Biometrics*, *12*, 51–56. R2

Ramachandran, K.V. (1958). A test of variances. *J. Amer. Stat. Assoc.*, *53*, 741–747. R2

Rankin, N.O. (1974). The harmonic mean method for one-way and two-way analysis of variance. *Biometrika*, *61*, 117–122. R4

Rao, C.R. and Kleffe, J. (1988). *Estimation of Variance Components and Applications*, North-Holland, Amsterdam. R2

Sahai, H. (1974). Simultaneous confidence intervals for variance components in some balanced random effects models. *Sankhyā*, Ser. B., 36, 278–287. R5,R6

Sahai, H. (1979). A bibliography on variance components. *Int. Stat. Rev.*, *47*, 177–222. R1,R2

Sahai, H. and Anderson, R.L. (1973). Confidence regions for variance ratios of random models for balanced data. *J. Amer. Stat. Assoc.*, *68*, 951–952. R5

Sahai, H., Khuri, A., and Kapadia, C.H. (1985). A second bibliography on variance components. *Comm. Stat. - Theor. Meth.*, *14*, 63–115. R1,R2

Samaranayake, V.A. and Bain, L.J. (1988). A confidence interval for treatment component-of-variance with applications to differences in means of two exponential distributions. *J. Stat. Comput. Simul.*, *29*, 317–332. R3

Satterthwaite, F.E. (1941). Synthesis of variance. *Psychometrika*, *6*, 309–316. R1,R3

Satterthwaite, F.E. (1946). An approximate distribution of estimates of variance components. *Biom. Bull.*, *2*, 110–114. R1,R3

Scheffé, H. (1956). Alternative models for the analysis of variance. *Ann. Math. Stat.*, *27*, 251–271. R7

Scheffé, H. (1959). *Analysis of Variance*, John Wiley & Sons, New York. R3,R5,R6,R7

Schulz, W. (1976). Sample size determination for the estimation of the variance when the underlying distribution is not normal. *Biom. Zeit.*, *18*, 535–546. R2

Searle, S.R. (1971). *Linear Models*, John Wiley & Sons, New York. R1,R2,R7

Searle, S.R. (1987). *Linear Models for Unbalanced Data*. John Wiley & Sons, New York. R6,R7

Searle, S.R. (1988). Mixed models and unbalanced data: Wherefrom, wherat, and whereto? *Comm. Stat.-Theor. Meth.*, *17*, 935–968. R1

Searle, S.R., Casella, G., and McCulloch, C.E. (1992). *Variance Components*, John Wiley & Sons, New York. R2,R7

Seely, J. (1980). Some remarks on exact confidence intervals for positive linear combinations of variance components. *J. Amer. Stat. Assoc.*, *75*, 372–374. R3

Seely, J.F. and El-Bassiouni, Y. (1983). Applying Wald's variance component test. *Ann. Stat.*, *11*, 197–201. R5,R6,RB

Seifert, B. (1979). Optimal testing for fixed effects in general balanced mixed classification models. *Statistics*, *10*, 237–256. R7

Seifert, B. (1981). Explicit formulae of exact tests in mixed balanced ANOVA-models. *Biometrical J.*, *23*, 535–550. R6,R7

*Sen, B., Graybill, F.A., and Ting, N. (1992). Confidence intervals on ratios of variance components for the unbalanced two factor nested model. Paper in review. R5

Singh, B. (1987). On the non-null distribution of ANOVA F-ratio in one way unbalanced random model. *Calcutta Stat. Assoc.*, *36*, 57–62. R4

Singhal, R.A. (1987). Confidence limits on heritability under nonnormal variations. *Biometrical J.*, *29*, 571–578. R4

Singhal, R.A., Tiwari, C.B., and Sahai, H. (1988). A selected and annotated bibliography on the robustness studies to non-normality in variance component models. *J. of the Jap. Stat. Soc.*, *18*, 195–206. R1

Smith, D.W. and Murray L.W. (1984). An alternative to Eisenhart's Model II and mixed model in the case of negative variance estimates. *J. Amer. Stat. Assoc.*, *79*, 145–151. R3,R4,R7

Smith, H.F. (1936). The problem of comparing the results of two experiments with unequal errors. *J. of the Counc. of Scient. and Indust. Res.*, *9*, 211–212. R1,R3

Snedecor, G.W. and Cochran, W.G. (1980). *Statistical Methods* (Seventh Edition), Iowa State University Press, Ames, Iowa. R4

Sokal, R. R., and Rohlf, F. J. (1969). *Biometry*, W.H. Freeman and Co., San Francisco. R5,R7

Solomon, P.J. (1985). Transformations for components of variance and covariance. *Biometrika*, *72*, 233–239. R1

Spjøtvoll, E. (1967). Optimum invariant tests in unbalanced variance components models. *Ann. Math. Stat.*, *38*, 422–428. R1,R4

Spjøtvoll, E. (1968). Confidence intervals and tests for variance ratios in unbalanced variance components models. *Rev. Internat. Stat. Inst.*, *36*, 37–42. R6

Srinivasan, S. (1986). Confidence intervals on functions of variance components in unbalanced two-way design models. Unpublished dissertation, Colorado State University, Ft. Collins, Colorado. R6

Srinivasan, S. and Graybill, F.A. (1991). Confidence intervals for proportions of total variation in unbalanced two-way components of variance models using unweighted means. *Comm. Stat.-Theor. Meth.*, *20*, 511–526. R6

Steel, R.G. and Torrie, J.H. (1960). *Principles and Procedures of Statistics*, McGraw-Hill, New York. R7

Swallow, W.H., and Searle, S.R. (1978). Minimum variance quadratic unbiased estimation (MIVQUE) of variance components. *Technometrics*, *20*, 265–272. R4

Tan, W.Y. (1981). The power function and an approximation for testing variance components in the presence of interaction in two-way random effects models. *Canad. J. Stat.*, *9*, 91–99. R6

Tan, W.Y. and Cheng, S.S. (1984). On testing variance components in three-stages unbalanced nested random effects models. *Sankhyā*, Series B, 46, 188–200. R5

Tan, W. Y., Tabatabai, M. A., and Balakrishnan, N. (1988). Harmonic mean approach to unbalanced random effects models under heteroscedasticity. *Comm. Stat.-Theor. Meth.*, *17(4)*, 1261–1286. R6

Tate, R.F. and Klett, G.W. (1959). Optimal confidence intervals for the variance of a normal distribution. *J. Amer. Stat. Assoc.*, *54*, 674–682. R2

Thomas, J.D. and Hultquist, R.A. (1978). Interval estimation for the unbalanced case of the one-way random effects model. *Ann. Stat.*, *6*, 582–587. R4

Thomsen, I.B. (1975). Testing hypotheses in unbalanced variance components models for two-way layouts. *Ann. Stat.*, *3*, 257–265. R6

Tietjen, G. L. (1974). Exact and approximate tests for unbalanced random effects designs. *Biometrics*, *30*, 573–581. R5

Ting, N., Burdick, R. K., Graybill, F.A., and Gui, R. (1989). One-sided confidence intervals on nonnegative sums of variance components. *Stat. and Prob. Lett.*, *8*, 129–135. R3

Ting, N., Burdick, R.K., Graybill, F.A., Jeyaratnam, S., and Lu, T.-F.C. (1990). Confidence intervals on linear combinations of variance components that are unrestricted in sign. *J. Stat. Comput. Simul.*, *35*, 135–143. R3

Ting, N., Burdick, R. K., and Graybill, F.A. (1991). Confidence intervals on ratios of positive linear combinations of variance components. *Stat. and Prob. Lett.*, *11*, 523–528. R3

*Ting, N. and Graybill, F.A. (1991). Approximate confidence interval on ratio of two variances in a two-way crossed model. Forthcoming in *Biometrical J.*, R3,R6

Tong, Y.L. (1979). Counterexamples to a result by Broemeling on simultaneous inferences for variance ratios of some mixed linear models. *Comm. Stat.-Theor. Meth.*, *8*, 1197–1204. R3

Tukey, J.W. (1951). Components in regression. *Biometrics*, *7*, 33–69. R3,R4

Venables, W. and James, A.T. (1978). Interval estimates for variance components. *Canad. J. Stat.*, *6*, 103–111. R1

Verdooren, L.R. (1988). Exact tests and confidence intervals for ratio of variance components in unbalanced two- and three-stage nested designs. *Comm. Stat. - Theor. Meth.*, *17*, 1197–1230. R4,R5

Vidmar, T.J. and Brunden, M.N. (1980). Optimal allocation with fixed power in a completely randomized design with levels of subsampling. *Comm. Stat. - Theor. Meth.*, *A9*, 757–763. R7

Wald, A. (1940). A note on the analysis of variance with unequal class frequencies. *Ann. Math. Stat.*, *11*, 96–100. R4,RB

Wald, A. (1941). On the analysis of variance in case of multiple classifications with unequal class frequencies. *Ann. Math. Stat.*, *12*, 346–350. RB

Wald, A. (1947). A note on regression analysis. *Ann. Math. Stat.*, *18*, 586–589. RB

Wang, C.M. (1978). Confidence intervals on functions of variance components. Unpublished dissertation, Colorado State University, Fort Collins, Colorado. R3

Wang, C.M. (1988a). β-expectation tolerance limits for balanced one-way random effects models. *Probability and Statistics: Essays in Honor of Franklin A. Graybill*, J.N. Srivastava (Ed.), North-Holland Publishing Co., 285–295. R1

Wang, C.M. (1988b). One-sided confidence intervals for the positive linear combination of two variances. *Comm. Stat.-Simula.*, *17(1)*, 283–292. R3

Wang, C.M. (1990a). On the lower bound of confidence coefficients for a confidence interval on variance components. *Biometrics*, *46*, 187–192. R3

Wang, C.M. (1990b). On ranges of confidence coefficients for confidence intervals on variance components. *Comm. Stat.-Simula.*, *19*, 1165–1178. R3

Wang, C.M. (1991). Approximate confidence intervals on positive linear combinations of expected mean squares. *Comm. Stat.-Simula.*, *20*, 81–96. R3

Wang, C.M. and Graybill, F.A. (1981). Confidence intervals on a ratio of variances in the two-factor nested components of variance model. *Comm. Statist. - Theor. Meth.*, *A10*, 1357–1368. R3,R5

Weeks, D. L. and Graybill, F. A. (1962). A minimal sufficient statistic for a general class of designs. *Sankhyā*, Ser. A, 24, 339–354. R6

Weir, J.A. (1949). Blood pH as a factor in genetic resistance to mouse typhoid. *J. of Infect. Dis.*, *84*, 252–274. R6

Welch, B.L. (1956). On linear combinations of several variances. *J. Amer. Stat. Assoc.*, *51*, 132–148. R1,R3

Westfall, P. (1988). Robustness and power of tests for a null variance ratio. *Biometrika*, *75*, 207–214. RB

Westfall, P. (1989). Power comparisons for invariant variance ratio tests in mixed ANOVA models. *Ann. of Stat.*, *17*, 318–326. RB

Wild, C.J. (1981). Interval estimates for variance components. *Canad. J. of Stat.*, *9*, 195–201. R1

Williams, J.S. (1962). A confidence interval for variance components. *Biometrika*, *49*, 278–281. R3, R4

Winer, B.J. (1971). *Statistical Principles in Experimental Design.* McGraw-Hill Book Co., New York. R6

Wonnacott, T. (1987). Confidence intervals or hypothesis tests? *J. of Appl. Stat.*, *14*, 195–201. R2

Yates, F. and Zacopanay, I. (1935). The estimation of the efficiency of sampling with special reference to sampling for yield in cereal experiments. *J. of Agric. Sci.*, *25*, 545–577. R1

Index